PROCEEDINGS OF THE 11TH INTERNATIONAL VETERINARY BEHAVIOUR MEETING

14th–16th September 2017, Samorin, Slovakia

PROCEEDINGS OF THE 11TH INTERNATIONAL VETERINARY BEHAVIOUR MEETING

14th–16th September 2017, Samorin, Slovakia

Edited by

Dr Sagi Denenberg, DVM, MRCVS

University of Bristol
School of Veterinary Sciences
Bristol, UK

CABI is a trading name of CAB International

CABI	CABI
Nosworthy Way	745 Atlantic Avenue
Wallingford	8th Floor
Oxfordshire OX10 8DE	Boston, MA 02111
UK	USA
Tel: +44 (0)1491 832111	Tel: +1 (617)682-9015
Fax: +44 (0)1491 833508	E-mail: cabi-nao@cabi.org
E-mail: info@cabi.org	
Website: www.cabi.org	

A catalogue record for this book is available from the British Library, London, UK.

ISBN-13: 978 1 78639 458 3

Commissioning editor: Jill Northcott
Production editor: Tracy Head

Typeset by SPi, Pondicherry, India.
Printed and bound in the UK by CPI Group (UK) Ltd, Croydon CR0 4YY.

Contents

POSTER PRESENTATIONS

Preface

It has been 20 years since Sarah Heath and Daniel Mills organised the first International Veterinary Behaviour Meeting (IVBM) in Birmingham, England. Since then, the meeting has taken place every 2 years around the globe including France, Canada, Australia, the USA, Italy, Scotland, South Africa, Portugal and Brazil.

Now, after two decades of successful meetings, it is my privilege to host this meeting, the 11th International Veterinary Behaviour Meeting, in Slovakia. Bringing the IVBM to Slovakia, and drawing large local interest, attest to the significance veterinary behaviour has within veterinary medicine. The field of veterinary behaviour is growing fast. With the help of meetings such as this, we carry the message to all of those who work with animals. The science presented in this meeting is at the forefront of our knowledge, and helps all of us to better ourselves in our profession.

Over the years the IVBM has grown and increasingly drawn people from different disciplines. Now we have not only veterinarians with interest in behavioural medicine, but also non-veterinary behaviourists, animal scientists, students and residents, animal welfare, law and ethics specialists, and industry representatives.

This meeting has several important partners, including the Slovak Institute of Neuroimmunology and the Slovak Society for Neuroscience. There are many people that I wish to thank for all their hard work – too many to mention all. This meeting would never have come to fruition without their help. Norbert Zilka was instrumental in contributing to the organisation of the IVBM. His hard and dedicated work included onsite arrangements, securing sponsorship, and technical details. I also wish to thank the Slovak Academy of Science and the University of Veterinary Medicine and Pharmacology in Kosice.

As part of the 11th IVBM, three other organisations hold their annual meeting. We are delighted to host the 7th annual meeting of the Animal Welfare, Science, Ethics and Law Veterinary Association (AWSELVA), the 23rd annual meeting of the European Society of Veterinary Clinical Ethology (ESVCE) and the

7th annual meeting of the European College of Animal Welfare and Behavioural Medicine (ECAWBM).

We had an amazing turnout of papers submitted for this conference. Over 100 papers were sent for review. To maintain high standards, quality, diversity and interest, we decided to accept the best ranking papers regardless of the author(s), country or species. Next, we decided to limit the number of presented papers to two for each author. My hope is that by following these guidelines, we have created an interesting, diverse and high-quality programme. Several authors were asked to present a poster instead of an oral presentation. Poster presenters also invested time and effort in preparing their posters, and their contribution should not be overlooked. Selected papers for this meeting represent speakers from all groups of interest in animal behaviour and welfare, arriving from all over the world. We worked hard to ensure that the language barrier is not going to affect this meeting. We have provided support to all those who needed grammatical, composition and spelling assistance. I am amazed, time and time again, by the efforts invested by non-native English authors who come and present their work.

At least three reviewers reviewed each paper and, where a disagreement arose, a fourth reviewer was tasked to provide a further critique. All papers that were selected underwent a rigorous evaluation. The reviewers did an amazing job within a short period. It is important for me to recognise these individuals who helped in this process, as their contribution is part of the success of this meeting. Some of these people volunteer each year to review papers, and I am indebted to them. The review committee consisted of Margaret Duxbury, Angelo Gazzano, Christine Halsberghe, Noa Harell, Anouck Haverbeke, Kim Kendall, Jacqui Ley, Anneli Muser Leyvraz, Melanie Rockman, Carlo Siracusa and Elizabeth Walsh. This group is truly an internationally flavoured committee.

This year we have invited four internationally acclaimed speakers to share their knowledge and expertise with us. I would like to thank Sarah Heath, Patrick Pageat, Barbara Sherman and Jo White, who have agreed to come and present during the meeting.

Finally, there are two more people to whom I am truly grateful for their advice, patience and calming effect even when, at times, the organisation seemed to collapse or problems arose. Gary Landsberg was there when I needed to ask a question, advice or assurances. This has been the case for over 15 years. My wife, Magdalena, suffered me through the organisation of this meeting, patiently listened and offered advice and tranquillity. For this, I am more than grateful.

Sagi Denenberg
Bristol, 2017

Platinum Sponsors

FELISCRATCH *by* FELIWAY®

FELIWAY® CLASSIC

FELIWAY® FRIENDS

Gold Sponsor

Bronze Sponsor

PURINA®
PRO PLAN®
VETERINARY
DIETS

Predicting Aggressive Behaviour: Which Factors Influence Biting and What is the Use of Temperament Tests?

BARBARA SCHOENING*

Practice for Behavioural Consultations, Hamburg, Germany

Conflict of interest: The author declares no conflict of interest.

Keywords: dog, aggressive behaviour, temperament tests, health status

Introduction

Dogs that have bitten are a common problem in behavioural counselling. Knowledge of the probability of an individual dog reacting aggressively, and which factors might promote or release a bite, are helpful for prevention, diagnosis and therapy rationale.

Materials and Methods

Over 16 years, 830 adult dogs of different breeds were tested in Hamburg, Germany, using a validated behavioural test for aggression (Schoening, 2006; NMEL, 2017). Dogs were scored between 1 (no aggression) and 6 (offensive biting without preceding threats). Breed, sex, bite history, medical history, obedience level, and negative-based or positive-based training were also recorded.

Results

Scores of 2 or higher were significantly associated with a history of biting (p=0.001). Sex of the dog and obedience level also influenced aggression scores

---

* bs@ethologin.de

and biting history. Intact males were more likely to bite other dogs (p=0.001). A history of biting was not significantly correlated to any particular means of training, but a poor obedience score was correlated with higher aggression scores (p<0.001). Owners using physical punishment had dogs with low obedience levels (p<0.001). There was a significant correlation between having bitten and a medical condition (p=0.005). The most frequent medical conditions were orthopaedic problems.

Discussion

These results show that formal and standardised tests can give some information on the probability of biting, and may be helpful in providing a prognosis and deciding on a therapy rationale. Factors such as obedience level and training method must be considered as well, and special emphasis should be placed on the medical condition of the dog.

References

NMEL (2017) Halten von Hunden. Available at: http://www.ml.niedersachsen.de/themen/tiergesundheit_tierschutz/tierschutz/halten-von-hunden-4745.html (accessed 29 June 2017) [in German].

Schoening, B. (2006) Evaluation and prediction of agonistic behaviour in the domestic dog. PhD thesis, University of Bristol, UK.

No Better Than Flipping a Coin: Reconsidering Canine Behaviour Evaluations in Animal Shelters*

GARY J. PATRONEK[1†] AND JANIS BRADLEY[2]

[1]*Center for Animals and Public Policy, Cummings School of Veterinary Medicine, Tufts University, North Grafton, Massachusetts, USA;* [2]*National Canine Research Council, Amenia, New York, USA*

Conflict of interest: Gary J. Patronek is a paid consultant to the National Canine Research Council, a subsidiary of Animal Farm Foundation. Janis Bradley is an employee of the National Canine Research Council.

Keywords: animal shelter, dog behaviour evaluation, aggression, dog personality, sensitivity, predictive value

Introduction

Our aim was to use existing data and established principles of diagnostic test evaluation to calculate the likelihood of reliably predicting problematically aggressive behaviours in adoptive homes among dogs residing in shelters based on exposing the dogs to a series of provocative stimuli (tests) in a semi-controlled environment (behaviour evaluations).

Materials and Methods

Robust values for sensitivity and specificity required for any valid diagnostic test have not been calculated for shelter canine behaviour evaluation, given the impracticality of adopting out all dogs testing positive for problematic behaviours, which would be necessary to collect unbiased follow up data. Therefore, the human literature on diagnostic test evaluation (both medical and behavioural)

---

* This paper has been published in full at: http://www.journalvetbehavior.com/article/S1558-7878(16)30069-7/pdf

† Corresponding author: gary.patronek@tufts.edu

©S. Denenberg 2017. *Proceedings of the 11th International Veterinary Behaviour Meeting* (ed. S. Denenberg)

was reviewed to derive the most optimistic and plausible estimates of sensitivity and specificity. Prevalence of the behaviours of interest (warning and biting behaviours deemed problematic by owners) was estimated based on a review of population-based studies of dog bite incidence and of reasons for rehoming/relinquishment among owners. These estimates were then used in standard simulations to calculate positive and negative predictive values for behaviour evaluations.

Results

For any plausible combination of sensitivity, specificity, and prevalence of biting and warning behaviours, a positive test result would be at best not much better than flipping a coin. The relatively low prevalence of the behaviours of interest in the population of shelter dogs who are evaluated results in unacceptable levels of false positive results using standardized behavioural tests.

Conclusion

Shelters already screen from adoption obviously dangerous dogs during the intake process. The authors suggest that instead of striving to bring out the worst in dogs in the stressful, transitional environment of a shelter and devoting scarce resources to inherently flawed formal evaluations that cannot increase public safety, simply interacting with dogs in ways that mirror what they are expected to do once adopted can both provide enrichment and help identify any additional dogs whose behaviour may be of concern.

Having Bitten is No One-way Ticket for Dogs: Rehabilitation Possibilities and Principles

BARBARA SCHOENING[1]*, LUISA FECHNER[2]
AND SUSANNE DAVID[3]

[1]Practice for Behavioural Consultations, Hamburg, Germany; [2]Horta da Valada, Estrada de Sta. Águeda, Portugal; [3]Hamburger Tierschutzverein von 1841 e.V., Hamburg, Germany

Conflict of interest: The authors declare no conflict of interest.

Keywords: dog, aggressive behaviour, rehabilitation training, temperament test

Introduction

Rehabilitation is the coordinated application of medical, social and pedagogical means influencing the physical and mental condition of an individual for the better. For dogs with a biting history, rehabilitation means to further the well-being of the dog and in parallel reducing the risk this dog poses for others.

Materials and Methods

In the last five years, 300 dogs with biting history were brought to the Hamburg humane shelter and underwent a special training, using positive, non-confrontational training methods. Following a review of the dogs medical and behavioural history, the dogs learned alternative behaviours for conflict situations via desensitisation and shaping processes. Training of bite inhibition, and improved tolerance levels for frustration, and inhibitory control complemented, when necessary. Training took between 6 and 12 months.

---

* Corresponding author: bs@ethologin.de

©S. Denenberg 2017. *Proceedings of the 11th International Veterinary Behaviour Meeting* (ed. S. Denenberg)

Results

Success was measured as either 'dog is adopted for life and is not reported with biting behaviour since', or 'dog is still living in the shelter (in group-housing) without biting behaviour reported'. The success rate was 87% for the first and 5% for the latter. Evaluation of 'before training' and 'after training' temperament tests underlined this success. There was a decrease in aggressive behaviour shown in the first and the second test (paired Wilcoxon signed rank test, p<0.02; Fechner 2016).

Discussion

The results show that individual and adequate training, following a thorough review of the medical and behavioural history, can improve the dog's behaviour and welfare and, at the same time, reduce the risk the dog poses for others.

References

Fechner, L. (2016) Assessing dogs using a temperament test – will training affect the outcome? A test-retest trial as a forensic tool. Dissertação de mestrado integrado em medicina veterinária. University of Lisbon, Veterinary Faculty, Lisbon.

Evaluation, Management and Welfare of Aggressive Shelter Dogs

MARIA CRISTINA OSELLA*

Research Institute in Semiochemistry and Applied Ethology (IRSEA), Quartier Salignan, France

Conflict of interest: The author declares no conflict of interest.

Keywords: dog, aggressive, shelter, evaluation, management, welfare

Disclaimer: This study was conducted according to current Italian animals ethics legislation.

Introduction

In Valle d'Aosta (Italy), a veterinary behavioural assistance was required for the management of aggressive dogs in the regional shelters. This project included the evaluation of 14 dogs, the guidelines for the new introductions, and the facility reorganisation. The aim was to reduce the animal and human risks and to increase the dogs' welfare.

Materials and Methods

Fourteen aggressive shelter dogs (males, 1–12 years old, different breeds) were housed after severe aggression and multiple bites incidences (Dehasse, 2002). All dogs were clinically and behaviourally assessed. The evaluation included phases of observation and physical approach based on tests reported in the literature (Michelazzi *et al.*, 2010). Rehoming was considered for each dog based on its behavioural profile and risk assessment. Appropriate management instructions to improve welfare parameters and human safety were defined, and a therapeutic approach was applied (cognitive and behavioural, and pharmacological when required).

* mc.osella@group-irsea.com

Results

Twelve months after the beginning of the project two dogs were successfully rehomed. No accidents were reported. The new guidelines were successfully applied in four new cases.

Discussion and Conclusions

Results of this study highlight the need for an accurate evaluation of the management and housing conditions for aggressive shelter dogs to improve their welfare. Shelters must outline detailed guidelines for procedures, safety tools and facilities to ensure appropriate prevention. Managing shelters in such a way should improve success rates of reintroducing or rehoming dogs and reduce the risk of future aggression.

Acknowledgements

The author acknowledges the public Veterinary Health Service (Azienda Unità Sanitaria Locale, Aoste, Italy) and A.VA.P.A. (Association Valdotaine pour la Protection des Animaux, Aoste, Italy) for their support in this study.

References

Dehasse, J. (2002) *Le Chien Aggressif.* Publibook, Paris, pp. 25–36.
Michelazzi, M., Levi, D., Fossati, P. and Scaglia, E. (2010) Evaluation of sheltered dangerous dogs. *Journal of Veterinary Behavior* 5, 35, DOI: http://dx.doi.org/10.1016/j.jveb.2009.10.019.

Keynote Presentation: Use of Psychopharmacology to Reduce Anxiety and Fear in Dogs and Cats: A Practical Approach

Barbara L. Sherman*

College of Veterinary Medicine, North Carolina State University, Raleigh, North Carolina, USA

Conflict of interest: The author has served on the Behaviour Advisory Boards for Elanco Animal Health, Lilly Animal Health, Novartis Animal Health, Shearing-Plough Animal Health, Virbac Animal Health and Zoetis Animal Health.

Funding: National Science Foundation (#557751)

Keywords: anxiety, fear, behavioural drugs, dogs, cats

Abstract

Anxiety and fear responses indicate impaired welfare in dogs and cats and often lead to erosion of the human–animal bond. In combination with simple behaviour modification regimes, psychopharmacologic agents may be given strategically to attenuate anxiety and fear responses (Hart and Cooper, 1996). The purpose of this review is to elucidate pharmacologic regimes that may be used strategically to manage such negative emotional states in dogs and cats, particularly in specific situations, such as travel, confinement and veterinary visits (Gruen *et al.*, in press; Mills and Simpson, 2002). The goal is for dogs and cats to be less anxious, less fearful, to have the capacity to learn new behavioural responses and to improve welfare. The application of specific drugs in a number of categories, including benzodiazepines, serotonin antagonist and reuptake inhibitors, and alpha 2 agonists, will be described, the literature briefly summarised, and case examples provided.

WHY USE Behavioural Drugs?

Behavioural drugs, also called psychopharmacologic agents, have important application to the treatment of behaviour problems encountered by the veterinary surgeon. In general, these drugs are most effectively used as an adjunctive

---

* barbara_sherman@ncsu.edu

to environmental management and behaviour modification regimes. In several clinical trials of behavioural problems in dogs and cats, enhanced treatment success was obtained with concurrent application of behavioural techniques and behavioural medication, compared to behavioural techniques alone. Behavioural drugs can:

- speed up the rate of improvement (when used with a behaviour plan);
- improve treatment success as measured by the number of animals that show significant improvement in clinical signs; and
- preserve the human–animal bond and retain the animal in the home.

Behavioural drugs may have general or specific effects. Some behavioural drugs, such as serotonergic agents, may decrease arousal or motivation that drive unacceptable behaviours, including hyperactivity, feline urine marking, and reactive aggression. The effects may sufficiently alter the problematic interactions between the pet and owner to allow environmental and behaviour modification programmes to be more effective. Behavioural drugs may also be used to treat specific anxieties and fears, such as separation anxiety and thunder phobia. Other agents are used to manage pathological disorders or organic states such as cognitive dysfunction or compulsive disorders. In summary, behavioural drugs are used to:

1. Treat conditions of high arousal or reactivity, such as:
- Hyperactivity
- Aggression
- Urine marking
2. Treat anxieties and fears, such as:
- Separation anxiety
- Thunder phobia
- Generalised anxiety
3. Treat abnormal behaviours, such as:
- Cognitive dysfunction
- Compulsive disorders

Behavioural drugs produce their effects through actions on neurotransmitters in the central nervous system. Especially important are the monoamine neurotransmitters serotonin (5-hydroxytryptamine or 5-HT), norepinephrine (NE; noradrenaline), and dopamine (DA), as well as acetylcholine (ACh), and gamma-aminobutyric acid (GABA). The potential for drug interactions must be considered when prescribing psychotropic drugs. For example, monoamine oxidase inhibitors, such as selegiline, when used concurrently with selective serotonin reuptake inhibitors, can lead to a serious medical condition called serotonin syndrome. A working knowledge of the specific agents and their potential interactions with other drugs is required (Stahl and Muntner, 2013).

WHY NOT Use Behavioural Drugs?

In general, behavioural drugs are well tolerated with routine concurrent medications, including routine heartworm and flea preventatives, antibiotics and anaesthetics. Some drug combinations should be avoided and are reasons NOT to use behavioural drugs. Concurrent use of monoamine oxidase inhibitors (including selegiline or amitraz) with tricyclic antidepressants or selective serotonin reuptake inhibitors should be avoided due to the risk of serotonin syndrome, a loss of physiological homeostatis due to an excess of serotonin in the CNS. Behavioural drugs should not be used in a number of situations, as when there are potential drug interactions, unresolved liability risks particularly in cases of aggression, and when pain or other confounding medical problems have been insufficiently evaluated, especially when physical, neurological or laboratory abnormalities exist. Due to their effects on the hepatic cytochrome P450 system, potential interactions with drugs classified as tricyclics or selective serotonin reuptake inhibitors include amitraz, cimetidine, and ketoconazole.

Avoid psychotropic drugs when there might be:

1. Potential drug interactions
2. Health risks
   - Physical/neuro/laboratory abnormalities, not resolved
   - PAIN, not evaluated and managed
3. Aggression (liability) risks, not safely managed

WHICH Drug to Use?

How does one choose a drug for behavioural therapy? The goal is to improve the behaviour without negative side effects. A good working knowledge of available therapeutic agents is a necessity (Schatzber and Nemeroff, 2017). This should include knowledge of the available drugs, their general behavioural actions, side effects (including potential drug interactions) and previously reported therapeutic effects (Tables 1 & 2). Understanding the neurochemistry of each drug helps to distinguish between agents likely to differ in their effects, so that if one drug is ineffective or poorly tolerated an alternative can be selected with different mechanisms of actions, side effect profile, or behavioural action. If available, an approved drug should be chosen first.

- Evaluate patient for medical problems/abnormalities (including pain).
- Collect a behavioural history.
- Make a behavioural diagnosis.
- Develop a management plan which includes behaviour modification and environmental management.
- Select an appropriate behavioural drug, if indicated.
- When possible, use an approved drug with clinical data in the species under treatment. To date, there are only few approved medications for dogs:
  - **Clomipramine** (King *et al.*, 2000), approved for treatment of canine separation anxiety, in combination with a behavioural plan, acts on serotonin and norepinephrine.
  - **Fluoxetine** (Wynchank and Berk, 1998; Pryor *et al.*, 2001; Simpson *et al.*, 2007; Landsberg *et al.*, 2008), approved for treatment of canine separation anxiety, in combination with a behavioural plan; acts on serotonin.

Table 1. Commonly used behavioural drugs for dogs.

Drug class	Drug name	PO dose & freq.	Side effects, comments
Phenothiazine	Acepromazine	0.1–2.2 mg/kg PRN storms	Tranquilliser, not an anti-anxiety agent
Benzodiazepine	Diazepam	0.5–2.2 mg/kg PRN storms	Rapidly metabolised, may inhibit learning
Benzodiazepine	Alprazolam	0.25–3.0 mg/dog PRN storms	Paradoxical excitation may occur
Benzodiazepine	Clorazepate	0.55–2.2 mg/kg q8-24h	Sedation, withdrawal syndrome if chronic use; requires an acid environment for absorption
Azaperone	Buspirone	1–2 mg/kg q12h	Mild GIT side effects (uncommon), changes in social behaviour may be evident
Tricyclic antidepressant	Amitriptyline	1–3 mg/kg 12h	Mild sedation, anticholinergic effects, mild GIT effects
Tricyclic antidepressant	Clomipramine	1–3 mg/kg q12h	Mild sedation, cardiac conduction disturbances in predisposed patients (humans)
Selective serotonin reuptake inhibitor (SSRI)	Fluoxetine	1–2 mg/kg q24h	Mild sedation or irritability, GIT especially inappetence
SSRI	Paroxetine	1–2 mg/kg q24h	Anticholinergic effects, restlessness
Atypical antidepressant	Trazodone	3–7 mg/kg PO PRN storms Maximum 300 mg/dose	Sedation, mild GIT side effects
Alpha-2 adrenergic agonist	Dexmedetomidine oromucosal gel	125 mcg/m^2 OTM	Sedation, lowers HR

- ○ **Selegiline** (Ruehl *et al.*, 1995), approved for treatment of canine cognitive dysfunction and chronic emotional disorders; acts on monoamine oxidase.
- ○ **Dexmedetomidine** (Korpivaara *et al.*, 2017), approved for treatment of noise aversion in dogs, acts on alpha 2 adrenergic receptors.

Psychotropic drugs may have behavioural effects beyond the specific application. For example, benzodiazepines and phenothiazines may inhibit the learning processes necessary for behaviour modification techniques. Alternately, tricyclic antidepressants may facilitate learning.

Drug classes

In the following sections, major anxiolytic and antidepressant drug classes are described and the drugs in most common use are discussed. Many other drugs, not discussed here, have behavioural effects and may be useful as part of a behavioural treatment programme. These include (to name a few) the antihistamines, antipsychotics, beta blockers, alpha agonists and opiate antagonists. In general,

Table 2. Commonly used behavioural drugs for cats.

Drug class	Drug name	PO dose & freq.	Side effects, comments
Phenothiazine	Acepromazine	0.5–1.1 mg/kg PRN	Tranquilliser, not a satisfactory anti-anxiety agent; paradoxical excitation
Benzodiazepine	Diazepam	0.2–0.5 mg/kg q12	Sedation, idiopathic hepatic necrosis (rare but may be fatal)
Azapirone	Buspirone	0.5–1.0 mg/kg q12-24h	Mild GIT side effects, changes in social behaviour (more confident, friendlier)
Tricyclic antidepressant	Amitriptyline	0.5–1.0 mg/kg q24h	Mild sedation, anticholinergic effects
Tricyclic antidepressant	Clomipramine	1–3 mg/kg q24h	Mild sedation, vomiting (if on empty stomach)
Selective serotonin reuptake inhibitor (SSRI)	Fluoxetine	0.5–1 mg/kg q24h	Mild sedation, inappetence
SSRI	Paroxetine	0.5 mg/kg q24h	Anticholinergic effects, sleepiness (mild)
Typical antidepressant	Trazodone	50 mg/cat PRN travel, vet visits	Mild sleepiness. Dose 2 hours before effect.

drugs within a group act on the same neurotransmitters and share a similar side effect profile. Behavioural drugs, listed below, produce their behavioural effects through actions on neurotransmitters and their receptors in the central nervous system. Especially important are the monoamine neurotransmitters serotonin (5-hydroxytryptamine or 5-HT; Mohammad-Zadeh *et al.*, 2008), norepinephrine (NE), and dopamine (DA), as well as acetylcholine (ACh), and gamma-aminobutyric acid (GABA).

I. Serotonergic drugs (antidepressants)
- Heterogeneous range of behavioural effects.
- Specific drugs differ in their effects on central neurotransmitters and their side effect profile.
- The various antidepressants thus differ in their effectiveness for treating a range of behavioural problems.
- May be individual differences in responses and effective doses.

A. TRICYCLIC ANTIDEPRESSANTS (TCA)
(King *et al.*, 2000; King *et al.*, 2004; Hart *et al.*, 2005).

1. Action: block the uptake of serotonin and norepinephrine.
2. Uses: control of aggression, urine spraying, excessive grooming, anxiety states, excessive vocalisation.
3. Side effects: mild anticholinergic side effects are common due to muscarinic blockade (dry mouth, urinary/faecal retention, reduced tear flow, mydriasis); antihistaminic, sedating (avoid by giving medication at bedtime). May lower

seizure threshold. May cause cardiac arrhythmias in predisposed humans, especially clomipramine. Side effects are not usually problematic in young, normal patients, but may be important considerations in animals with medical complications.
4. Examples: Clomipramine, Amitriptyline, Imipramine.

B. SELECTIVE SEROTONIN REUPTAKE INHIBITORS (SSRI)
1. Action: serotonin reuptake blocker.
2. Uses: treatment for refractory urine spraying, compulsive disorders, anxiety states, aggression.
3. Side effects: gastrointestinal signs (anorexia, nausea, diarrhoea) since most of the serotonin receptors in the body are found in the digestive system. Side effects (up to 25% in humans) may be avoided by starting at a low dose for 1 week, then increasing. Other side effects include anxiety, irritability and insomnia.
4. Examples: Fluoxetine, Sertraline, Paroxetine, Citalopram, Escitalopram.

C. SEROTONIN ANTAGONIST REUPTAKE INHIBITORS (SARI)
(Gruen and Sherman, 2008; Gruen *et al.*, 2014; Stevens *et al.*, 2016).

1. Action: serotonin antagonist and reuptake inhibitor.
2. Uses: reduce arousal and anxiety in dogs and cats; may be mildly sedating.
3. Side effects: lethargy, gastrointestinal signs, paradoxical excitation (rare).
4. Example: Trazodone.

II. Anxiolytics
The anxiolytics include benzodiazepines, azapirones, barbiturates, antihistamines and alpha-2 agonists. Only benzodiazepines, azapirones and alpha-2 agonists are considered below. Other drugs, such as antidepressants have anti-anxiety properties and are clinically useful in the treatment of anxiety states.

A. BENZODIAZEPINES
(Center *et al.*,1996; Herron *et al.*, 2008).

1. Action: activate benzodiazepine receptors; facilitate GABA (inhibitory neurotransmitter) in the CNS.
2. Uses: fears and phobias, states of anxiety and arousal in dogs. Especially useful for treatment of anxiety and arousal, as thunderstorm phobia. May be used in combination with TCAs and SSRIs.
3. Side effects: sedation, ataxia, muscle relaxation, increased appetite, paradoxical excitation, memory deficits, idiopathic hepatic necrosis in cats (rare, but may be fatal), discontinuation reactions.
4. Examples: Diazepam, Alprazolam, Lorazepam, Clorazepate.

B. AZAPIRONES
(Hart *et al.*, 1993; Mills and Ledger, 2001; Mealey *et al.*, 2004).

1. Action: serotonergic and dopaminergic mechanisms.
2. Uses: numerous, useful for urine spraying (improvement in 55% of cats), anxiety states. Ineffective for the control of panic disorder in humans or for the panic-like states of canine separation anxiety or thunderstorm phobia.

3. Side effects: gastrointestinal signs (uncommon), irritability, changes in social relationships.
4. Example: Buspirone.

C. ALPHA-2 ADRENERGIC AGONISTS
(Ogata and Dodman, 2011; Hopfensperger *et al.*, 2013; Korpivaara *et al.*, 2017).

1. Action: inhibits release of noradrenaline from noradrenergic neurons.
2. Uses: noise aversions, situational anxiety such as veterinary visits.
3. Side effects: bradycardia, sedation, paradoxical excitation, emesis (cats).
4. Example: Dexmedetomidine, Clonidine.

III. Monoamine oxidase inhibitors
(Ruehl *et al.*, 1995).

1. Action: inhibition of monoamine oxidase A or B enzymes.
2. Uses: cognitive dysfunction and emotional disorders.
3. Side effects: emesis, restlessness, anorexia, lethargy, repetitive movements.
4. Example: Selegiline.

Side effects of behavioural drugs

Although usually mild, side effects are observed with behavioural drugs. In some cases, side effects of specific drugs may contraindicate their use in specific individuals. For example, anticholinergic side effects are commonly associated with tricyclic antidepressants, due to muscarinic blockade. Dry mouth, constipation, urinary retention, reduced tear flow, and mydriasis may be noted. Although not usually problematic in young, normal patients, they are important considerations in animals with medical complications. Because of their potential effects on cardiac conduction, tricyclics should not be used in patients diagnosed with cardiac conduction disturbances. Because of the potential for the rare condition, feline idiopathic hepatic necrosis, oral benzodiazepines should not be used in cats to treat behavioural problems, except as a last resort.

Clinical Treatment of Behavioural Problems

After medical aetiologies have been ruled out, behavioural treatment consists of environmental management, behaviour modification techniques and behavioural medications (if indicated).

What to expect?

Many behavioural drugs require 1–3 weeks to initial behavioural effects; maximum effect may take more time. Clients should be counselled to expect a latency to effect and to be aware that side effects may occur immediately. It is important to note that a behavioural plan that reduces arousal and anxiety should be implemented prior to or concurrent with the administration of behavioural medication. In some cases, a first-choice behavioural drug may be insufficiently effective. Another agent may be chosen to optimise treatment outcome. Clients should be advised of this possibility at the onset of treatment.

- Gradual improvement in target signs for many animals.
- Best effect if used with behavioural management.
- Behavioural plan simplified:
  - Make a list of problem situations so that they can be avoided.
  - Use safe places, leashes, halters, gates to AVOID problem situations and clear instructions and rewards to teach new, acceptable responses.
  - Avoid punishment.

How to enhance therapy?

To simplify treatment, single drug therapy should be attempted first, followed by dose manipulation. However, improved management of some canine anxiety disorders may be obtained with the use of rational drug combinations. Tricyclic antidepressants and benzodiazepines may be used together since they each affect completely different neurotransmitter systems (Crowell-Davis *et al.*, 2003). Tricyclic antidepressants and selective serotonin reuptake inhibitors may be used with buspirone or trazodone (Gruen and Sherman, 2008). Benzodiazepines may be used with monoamine oxidase inhibitors. Adjunctive agents which can be used with the baseline drugs:

- Benzodiazepines
  - Alprazolam
  - Lorazepam
  - Clorazepate
- Atypical agents (not used with selegiline)
  - Buspirone
  - Trazodone

When drug combinations are used, an understanding of relevant neurotransmitters is especially important in order to avoid adverse drug reactions. Some drug combinations should be avoided. The concurrent use of monoamine oxidase inhibitors [including selegiline (L-deprenil) or amitraz] with tricyclic antidepressants or selective serotonin reuptake inhibitors should be avoided due to the risk of serotonin syndrome.

Medication management

Successful treatment of behavioural problems requires a schedule of follow-ups with the client, in person or by telephone. If problematic side effects or inefficacy occur, the dose can be adjusted. If a medication is not effective after dose adjustment, selection of an agent from another drug class is recommended. If a drug in one class is not sufficiently effective, and appropriate dose adjustments have been made, that drug should be discontinued and a drug from another drug class should be prescribed, according to the 'wash-out' instructions for the specific agents. In some cases, more than one drug is used concurrently, although this practice requires knowledge of the available drugs, their general behavioural actions, side effects (including potential drug interactions) and previously reported therapeutic effects. Understanding the neurochemistry of each drug helps to distinguish between agents likely to differ in their effects, so that if one drug is ineffective or poorly tolerated an alternative can be selected with different mechanisms of actions, side effect profile or behavioural action.

Conclusions

Behavioural pharmacotherapy is an important component of veterinary behaviour therapy.

Psychoactive drugs are most effective when used as part of a comprehensive programme involving behaviour modification and environmental management.

In general, behavioural drugs are well tolerated, although knowledge of the action and potential side effects of these agents is necessary to guide the practitioner in their use.

References

Center, S.A., Eston, T.H., Rowland, P.H., Rosen, D.K., Reitz, B.L., Brunt, J.E., Rodan, I., House, J., Banks, S., Lynch, L.R., Dring, L.A. and Levy, J.K. (1996) Fulminant hepatic failure associated with oral administration of diazepam in 12 cats. *Journal of American Veterinary Medical Association* 209, 618–625.

Crowell-Davis, S.L., Seibert, L.M., Sung, W., Parthasarathy, V. and Curtis, T.M. (2003) Use of clomipramine, alprazolam, and behavior modification for treatment of storm phobia in dogs. *Journal of American Veterinary Medical Association* 222, 744–748.

Gruen, M.E. and Sherman, B.L. (2008) Use of trazodone as an adjunctive agent in the treatment of canine anxiety disorders: 56 cases (1995–2007). *Journal of American Veterinary Medical Association* 233, 1902–1907.

Gruen, M.E., Roe, S., Griffith, E., Hamilton, A. and Sherman, B.L. (2014) The use of trazodone to facilitate post-surgical confinement in dogs. *Journal of American Veterinary Medical Association* 245, 296–301.

Gruen, M.E., Sherman, B.L. and Papich, M. (In press) Drugs affecting animal behavior. In: Papich, M.G. and Rivier, J.E. (eds) *Veterinary Pharmacology and Therapeutics*. 10th edn, Ames IA, Wiley.

Hart, B.L. and Cooper, L.L. (1996) Integrating use of psychotropic drugs with environmental management and behavioral modification for treatment of problem behavior in animals. *Journal of American Veterinary Medical Association* 209, 1549–1551.

Hart, B.L., Eckstein, R.A., Powell, K.L. and Dodman, N.H. (1993) Effectiveness of buspirone on urine spraying and inappropriate urination in cats. *Journal of American Veterinary Medical Association* 203, 254–258.

Hart, B.L., Cliff, K.D., Tynes, V.V. and Bergman, L. (2005) Control of urine marking by use of long-term treatment with fluoxetine or clomipramine in cats. *Journal of American Veterinary Medical Association* 26, 378–382.

Herron, M.E., Shofer, F.S. and Reisner, I.R. (2008) Retrospective evaluation of the effects of diazepam in dogs with anxiety-related behavior problems. *Journal of American Veterinary Medical Association* 233, 1420–1424.

Hopfensperger, M.J., Messinger, K.M., Papich, M.G. and Sherman, B.L. (2013) The use of oral transmucosal detomidine hydrochloride gel to facilitate handling in dogs. *Journal of Veterinary Behavior* 9, 114–123.

King, J.N., Simpson, B.S., Overall, K.L., Appleby, D., Pageat, P., Ross, C., Chaurand, J.P., Heath, S., Beata, C., Weiss, A.B., Muller, G., Paris, T., Bataille, B.G., Parker, J., Petit, S. and Wren, J. (2000) Treatment of separation anxiety in dogs with clomipramine: results from a prospective, ramdomized, double-blind, placebo-controlled, parallel-group, multicenter clinical trial. *Applied Animal Behavior Science* 67, 255–275.

King, J.N., Steffan, J., Heath, S.E., Simpson, B.S., Crowell-Davis, S.L., Harrington, L.J., Weiss, A.B. and Seewald, W. (2004) Determination of the dosage of clomipramine for the treatment of urine spraying in cats. *Journal of American Veterinary Medical Association* 225, 881–887.

Korpivaara, M., Laapas, K., Huhtinen, M., Schoning, B. and Overall, K. (2017) Dexmedetomidine oromucosal gel for noise-associated acute anxiety and fear in dogs—a randomised, double-blind, placebo-controlled clinical study. *Veterinary Record* 180, 365, DOI: 10.1136/vr.104045.

Landsberg, G.M., Melese, P., Sherman, B.L., Neilson, J.C., Zimmerman, A. and Clarke, T.P. (2008) Effectiveness of fluoxetine chewable tablets in the treatment of canine separation anxiety. *Journal of Veterinary Behavior* 3, 12–19.

Mealey, K.L., Peck, K.E., Bennett, B.S., Sellon, R.K., Swinney, G.R., Melzer, K., Gokhale, S.A. and Drone, T.M. (2004) Systemic absorption of amitriptyline and buspirone after oral and transdermal administration to healthy cats. *Journal of Veterinary Internal Medicine* 18, 43–46.

Mills, D.S. and Ledger, R. (2001) The effect of oral selegiline hydrochloride on learning and training in the dog: a psychobiological interpretation. *Progress Neuro-psychopharmacology and Biology Psychiatry* 25, 1597–1613.

Mills, D. and Simpson, B.S. (2002) Psychotropic agents. In: Horwitz, D., Mills, D. and Heath, S. (eds) *British Small Animal Veterinary Association Manual of Canine and Feline Behavioural Medicine.* BSAVA Publishing, Gloucester, UK, pp. 237–248.

Mohammad-Zadeh, L.F., Moses, L. and Gwaltney-Brant, S.M. (2008) Serotonin: a review. *Journal of Veterinary Pharmacology and Therapeutics* 31, 187–199.

Ogata, N. and Dodman, N. (2011) The use of clonidine in the treatment of fear-based behavior problems in dogs: an open trial. *Journal of Veterinary Behavior* 6, 130–137.

Pryor, P.A., Hart, B.L., Cliff, K.D. and Bain, M.J. (2001) Effects of a selective serotonin reuptake inhibitor on urine spraying behavior in cats. *Journal of American Veterinary Medical Association* 219, 1557–1561.

Ruehl, W.W., Bruyette, D.S., DePaoli, A., Cotman, C.W., Milgram, N.W. and Cummings, B.J. (1995) Canine cognitive dysfunction as a model for human age-related cognitive decline, dementia, and Alzheimer's disease: clinical presentation, cognitive testing, pathology, and response to L-deprenyl therapy. *Progress in Brain Research* 106, 217–225.

Schatzberg, A.F. and Nemeroff, C.B. (eds) (2017) *The American Psychiatric Association Publishing Textbook of Psychopharmacology.* 5th edn. American Psychiatric Association Press, Washington, DC.

Simpson, B.S., Landsberg, G.M., Reisner, I.R., Ciribassi, J.J., Horwitz, D., Houpt, K.A., Kroll, T.L., Luescher, A., Moffat, K.S., Douglass, G., Roberston-Plouch, C., Veenhuizen, M.F., Zimmerman, A. and Clarke, T.P. (2007) Effects of reconcile (Fluoxetine) chewable tablets plus behavior management for canine separation anxiety. *Veterinary Therapeutics* 8, 18–31.

Stahl, S.M. and Muntner, N. (2013) *Stahl's Essential Psychopharmacology*, 4th edn. Cambridge University Press, Cambridge, UK.

Stevens, B.J., Frantz, E., Orlando, J.M., Griffitth, E., Harden, L.B., Gruen, M.E. and Sherman, B.L. (2016) The use of trazodone to reduce feline travel anxiety and improve veterinary exam tractability. *Journal of American Veterinary Medical Association* 249, 202–207.

Wynchank, D. and Berk, M. (1998) Fluoxetine treatment of acral lick dermatitis in dogs: a placebo-controlled randomized double blind trial. *Depression and Anxiety* 8, 21–23.

Preventing Travel Anxiety Using Dexmedetomidine Hydrochloride Oromucosal Gel

Marta Amat[1]*, Susana Le Brech[1], Camino García-Morato[1], Déborah Temple[1], Marta Salichs[2], Bibiana Prades[2], Tomàs Camps[1] and Xavier Manteca[1]

[1]School of Veterinary Medicine (Autonomous University of Barcelona), Barcelona, Spain; [2]ECUPHAR Veterinaria S.L.U., Barcelona, Spain

Conflict of interest: Marta Salichs and Bibiana Prades work for Ecuphar.

Keywords: dog, travel anxiety, dexmedetomidine hydrochloride, oromucosal gel

Introduction

Dogs frequently show anxiety when travelling by car. The aim of this study was to evaluate the efficacy of dexmedetomidine hydrochloride 0.1 mg/ml oromucosal gel (Sileo®, Ecuphar/Orion), an alpha-2 agonist, to reduce anxiety in dogs during transportation by car.

Material and Methods

A triple blind crossover design was used. Twelve beagle dogs were included in the study. The duration of the transport by car was 10 minutes, and the route was the same for all dogs. Each dog was subjected to two test situations: treatment and control. Two hours before the treatment phase each dog received a dose of Sileo® (125 µg/m^2) but not before the control phase. Behavioural observations were made by focal and scan sampling. Observations included the following variables: urinating (yes/no), defaecating (yes/no), vomiting (yes/no), salivating (yes/no), panting (yes/no), attempts to escape (yes/no), yawning (yes/no), sniffing

---

* Corresponding author: Marta.amat@uab.es

©S. Denenberg 2017. *Proceedings of the 11th International Veterinary Behaviour Meeting* (ed. S. Denenberg)

(number of events) and lip/nose licking (number of events). The posture of the dog (standing, sitting, laying down) was also recorded. Data were analysed using non-parametric generalised estimating equations, and the dog was included as a repeated variable.

Results

No dogs urinated, defaecated or vomited in either trial phases. The frequency of panting was significantly (p=0.008) lower in the treatment group as compared to control (25% and 83% respectively). Similarly, the frequency of yawning was significantly (p=0.05) lower in the treatment group (42% vs 67%). The other behavioural variables were not significantly different between treatments.

Conclusions

These results suggest that Sileo® can be useful to reduce anxiety in dogs during transportation by car.

Oromucosal Dexmedetomidine Gel for Alleviation of Fear and Anxiety in Dogs During Minor Veterinary or Husbandry Procedures

Mira Korpivaara[1]*, Mirja Huhtinen[1], John Aspegren[1] and Karen Overall[2]

[1]Orion Corporation, Research and Development, Turku, Finland; [2]Biology Department, University of Pennsylvania, Philadelphia, Pennsylvania, USA

Conflict of interest: Mira Korpivaara, Mirja Huhtinen and John Aspegren are employees of Orion Corporation Orion Pharma Finland. Karen Overall was a paid consultant for this study. Orion Corporation funded the study.

Keywords: behaviour, oromucosal dexmedetomidine, veterinary anxiety

Introduction

The primary objective was to evaluate the efficacy of dexmedetomidine gel in improving the ability to perform a physical examination and a minor veterinary or husbandry procedure in dogs suffering from fear at the veterinary surgery.

Material and Methods

Seventy-four client-owned dogs were enrolled into a randomised, double blinded, placebo-controlled study. Eligibility of dogs was confirmed at a baseline visit. This was a multicentre, dose-titration study. Two dexmedetomidine gel doses ($125 \ \mu g/m^2$ and $250 \ \mu g/m^2$) were compared to placebo for efficacy and safety. Investigators assessed the ability to perform the intended procedure at the treatment visit using a scale from 1 (procedure could be easily performed) to 5 (not possible). The distributions of the five scores were compared between dexmedetomidine and placebo treatments using a generalised linear model for ordinal data.

* Corresponding author: mira.korpivaara@orionpharma.com

Results

The treatment effect was statistically significant (p=0.0136), and the ability to perform the procedure was better with both 125 µg/m^2 (OR 4.9; 90% CI 1.9–13.1; p=0.0072) and 250 µg/m^2 (OR 3.5; 90% CI 1.4–8.9; p=0.0278) doses of dexmedetomidine compared with placebo. No sedation was noted. A decrease of 29–30 bpm in heart rate from the initial high values was seen in dexmedetomidine-treated dogs, but not for those on placebo, supporting the physiological anxiety-lowering effect of the study treatment.

Conclusion

Dexmedetomidine gel alleviated fear and anxiety during minor veterinary or husbandry procedures in dogs previously reported to suffer from fear and anxiety during veterinary visits. Both dexmedetomidine gel doses 125 µg/m^2 and 250 µg/m^2 were effective and no safety concerns were observed.

Avalanche Dogs Can Locate 'Buried Victims' by Perceiving the Human Breath Under the Snow

Silvana Diverio[1]*, Laura Menchetti[1], Martina Iaboni[1], Giacomo Riggio[2], Costanza Azzari[3], Anselmo Cagnati[4], Walter Di Mari[5] and Michele Matteo Santoro[6]

[1]Laboratory of Ethology and Animal Welfare (LEBA) Department of Veterinary Medicine, Perugia University, Italy; [2]Veterinary Consultant, Rome, Italy; [3]Veterinary Consultant, Turin, Italy; [4]ARPA (Veneto Regional Agency for the Environment Protection), Arabba, Italy; [5]GdF (Military Force of Guardia di Finanza), Direzione Veterinaria e Cinofili, Rome, Italy; [6]SAGF - Alpine Rescue of Guardia di Finanza, Predazzo, Italy

Conflict of interest: The authors declare no conflict of interest.

Keywords: behaviour, working dog, human scent, body postures, avalanche

Introduction

Which components of the human scent steer dogs in finding a buried victim? The aim of this study was to evaluate if rescue dogs can locate a 'buried victim' only by perceiving the human's breath under a layer of snow in an avalanche simulation context, and which dog's body postures are associated with a successful search.

Material and Methods

An operator blew for 18 min inside a tube buried under 1 m of snow on a limited field (25 m × 25 m). After a 20-min break, the Alpine Rescue of Guardia di Finanza (SAGF) Units (n=15) had to find a 'buried victim', i.e. the breath coming out from the hidden tube (15-min period). When the SAGF dogs signalled the

---

* Corresponding author: silvana.diverio@unipg.it

human breath and found the end part of the buried tube, the trial was considered successful. The protocol was replicated.

Results

SAGF-Units performed successfully most of the trials (26/30, 87%; $p<0.0001$) within a median latency time of 63 s (IQR: 39–201 s). Dog's gender, age or breed and trial's number did not affect success rate. During the search, dogs spent most of the time ($p<0.0001$) with the head low (78%) and facing down (79%), neutral posture (91%), back parallel tail (50%), ears forward (74%) and closed mouth (64%). On successful trials, dogs spent more time with a high posture ($p<0.001$), head low ($p<0.05$) and facing down ($p<0.01$), vertical tail ($p<0.01$), and ears forward ($p<0.05$) compared with failed trials.

Conclusion

In conclusion, human breath scent proved to be a valuable indicator to locate a 'buried victim' for avalanche dogs.

Responses of Anxious Dogs to a Simple Behaviour Modification Protocol While Waiting in a Veterinary Hospital

Barbara Sherman[1]*, Jalika Joyner[1], Sherrie Yuschak[1], Katherine Walker[2], Justin Kuhn[2], John Majikes[2], Hongyu Ru[2], Sean Mealin[2], Rita Brugarolas[2], David Roberts[2] and Alper Bozkurt[2]

[1]College of Veterinary Medicine, North Carolina State University, Raleigh, North Carolina, USA; [2]College of Engineering, North Carolina State University, Raleigh, North Carolina, USA

Funding: The study was funded by the National Science Foundation (#557751), the NC State Veterinary Scholars Program and Fund for Discovery.

Conflict of interest: The authors declare no conflict of interest.

Keywords: anxiety, behaviour, dog, heart rate, panting

Introduction

Signs of anxiety are commonly demonstrated by dogs awaiting care at a veterinary facility. Our hypothesis was that a simple owner-implemented behaviour modification protocol would attenuate behavioural and physiological signs of anxiety compared to untreated controls.

Methods

Dogs with a history of anxious behaviour in a veterinary hospital were recruited and randomly assigned to treatment or control groups. Each dog was instrumented

---

* Corresponding author: barbara_sherman@ncsu.edu

with a harness that collected heart rate and panting rate for 21 minutes. For treatment, the owner implemented a simple positive behavioural modification protocol. For control, the owner benignly ignored the dog. A video camera continuously recorded the dog's behaviour. For analysis, recordings were divided into seven 3-minute segments, the first defined as 'baseline'. Then each dog's behaviour during each subsequent segment was compared to baseline by a trained observer masked to segment order. The score assigned to each segment reflected the duration and intensity of anxiety-associated behaviours on negative or positive axes. Physiological data consisted of mean heart rate and percentage of time spent panting.

Results

Thirteen dogs completed the study. The results revealed that all dogs were anxious at the start of the experiment, most expressing 'negative' global anxiety signs. There were individual responses to treatment over time, with some dogs exhibiting lower heart rate, panting and anxiety scores.

Conclusion

The findings suggest that physiological measures may augment our appreciation of anxiety in canine patients. A simple behavioural protocol may be implemented by owners to reduce dog anxiety while waiting in a veterinary facility.

Link Between Chronic Gastric Diseases and Anxiety in Dogs

MURIEL MARION[1]*, PATRICK LECOINDRE[2], NATHALIE MARLOIS[3], CATHERINE MÈGE[4], CLAUDE BÉATA[5], GUILLAUME SARCEY[6] AND GÉRARD MULLER[7]

[1]Cabinet médico-chirurgical Montolivet, Marseille, France; [2]CVC Clinique Vétérinaire des Cerisioz, St Priest, France; [3]Clinique Vétérinaire de l'Albarine, Ambérieu en Bugey, France; [4]Clinique Vétérinaire Les Grands Crus, Chenôve, France; [5]Consultant Vétérinaire, Toulon, France; [6]Clinique vétérinaire Saint Roch, Gap, France; [7]Clinique Vétérinaire de Lille St Maurice, Lille, France

Conflict of interest: The authors declare no conflict of interest.

Keywords: chronic gastric disease, anxiety, dog

Introduction

Anxiety in dogs manifests as a collection of physical and behavioural signs. The clinical signs that are often reported include trembling, panting, urination and defaecation (Overall *et al.*, 2001; Tiira *et al.*, 2016). Aggressiveness, destructive behaviour, wandering, running away, inhibition and vocalising are some of the frequently reported behaviours.

The differential diagnosis of chronic gastric disease in dogs includes digestive and systemic causes. The area postrema, the vomiting centre, is in the medulla oblongata and includes many nerve ending and tracts that can lead to emesis. Elwood *et al.* (2010) carried out a systematic review of the causes of emesis in dogs. This review included gastrointestinal diseases such as gastric ulceration, pyloric stenosis, infection, inflammatory bowel diseases, gastric and intestinal neoplasia, and obstructions or intestinal occlusions. Vomiting may also occur as a result of other conditions such as hepatobiliary or pancreatic disease. Finally,

---

* Corresponding author: muriel.marion@free.fr

there are systemic diseases including metabolic changes, and toxic or drug-induced causes may lead to vomiting.

In human medicine, the link between anxiety and chronic gastric disease has been the subject of many studies. Out of 1641 patients with gastrointestinal complaints, 84.1% (1379) also suffered from anxiety (Addolorato *et al.*, 2008). A prospective study of 4181 adults reported that individuals with gastritis exhibit significantly more anxiety than the rest of the population (Goodwin *et al.*, 2013).

In canine medicine, only a few studies have investigated the behavioural aspect of gastric illness. Some studies have investigated a possible link between functional digestive disorders and anxiety. Dogs with chronic idiopathic large bowel disease (CILBD) had significantly higher anxiety scores than control dogs (Reiwald *et al.*, 2013). Another study showed that more than 9% (8/85) of dogs with CILBD improved when treated with a psychotropic agent (Lecoindre and Gaschen, 2011). These studies addressed intestinal disorders; the work reported here involves solely functional disorders of gastric origin.

To determine whether there is a potential link between chronic gastric disorders and behavioural disorders in dogs, we evaluated anxiety in a population of dogs with chronic gastric disease compared to control dogs.

Materials and Methods

Twenty dogs with chronic gastric disorders were compared to 20 healthy dogs. The presence of *Helicobacter* spp. was not a criterion for exclusion. The inclusion criteria for the ill animals were as follows:

1. History of chronic gastric disorders, accompanied by vomiting or dyspepsia, without associated diarrhoea (intermittent or chronic). Dyspepsia was defined by the postprandial appearance of signs of digestive discomfort. A dog was considered to have dyspepsia if one of the following three signs was present: gastric distension, abdominal pain or gastroesophageal reflux.
2. Absence of comorbidities, excluding any behavioural disorder.
3. Normal blood biochemistry.
4. Absence of macroscopic lesions upon endoscopic examination.
5. Absence of histological lesions on biopsies taken during the endoscopic examination.
6. Ineffectiveness of treatments or diets in preventing recurrence.
7. All selected animals were under medical care and are on an easily digestible diet.

The control group included dogs without a digestive disease. Each control dog was paired with a matching ill dog. The age of each control dog was within 10 months (higher or lower) of the matching ill dog. The control dog and ill dog were required to be of the same breed (based on phenotype). Both dogs in the pair were the same sex and reproductive status.

Clinical, biochemical and endoscopic examinations were performed on each of the dogs with chronic gastric disease by the same veterinarian specialist (in internal medicine). The BIOMNIS laboratory performed histopathology. The clinical examination of each control dog was carried out by its regular veterinarian. All physical examinations and additional tests were done with the owner's consent, and the emphasis on reducing the risk of pain, injury and fear.

The Evaluation of a Dog's Emotional Disorders (EDED) scale is used to score anxiety in dogs (Table 1) (Pageat, 1990; Reiwald *et al.*, 2013). A table is completed for each dog based on an interview with the owners (Fig. 1). All interviews were conducted by the same veterinarian, who is experienced in the behavioural evaluation and trained in administering the questionnaire. A dog is considered to have anxiety if its score is greater than or equal to 17 and less than or equal to 35.

The results obtained for each group were compared using Student test and Wilcoxon test.

The statistical calculations were performed using R software. Probability values less than 0.05 were considered significant.

Results

The average age of the dogs included in the study was 5.7 years, and 75% were male (15/20) (Fig. 2). Yorkshire Terriers, Labrador Retrievers, and Standard Poodles were the most common breeds. Smaller breeds were more represented than large or medium breeds, but the small sample size did not allow us to identify any significant difference.

In the group of dogs with chronic gastric disease, 85% (17/20) had an EDED score between 17 and 35 (inclusive), three dogs scored less than 17, and no dogs scored over 35 (Table 2). Only one of the control dogs (5%) had an EDED score great than 17, and could thus be classified as having anxiety (Fig. 3). The ill animals primarily exhibited vomiting (75%). The average EDED score of dogs with dyspepsia was 23.2, compared to 19.6 for dogs who exhibited vomiting. There was a significant difference (Wilcoxon, p=0.00023) between the groups regarding the EDED results; ill dogs scored higher. The average EDED score in the ill dog group was 20.5, and the median was 20.5 with a variance of 21.8. In the control group, the average EDED score was 11.5, the median was 11, and the variance was 6.25. There was no significant difference between the scores obtained for the dogs with dyspepsia and those who exhibited vomiting (Student test, p=0.28).

Discussion

The presence of *Helicobacter* spp. in the stomach of ill dogs was not used as a criterion for exclusion in this study. In humans, gastric ulcers associated with dyspepsia can have an infectious aetiology. *Helicobacter pylori* is identified in 30% of healthy individuals and 90% of patients with gastric ulcers (Heams, 1996).

Table 1. Evaluation of a Dog's Emotional Disorder (EDED) scale reference values.

EDED Score	9 to 12	13 to 16	17 to 35	36 to 44
Status	Normal	Phobic	Anxious	Mood disorder

Behaviours		Item	V1	V2	V3	V4
SELF FOCUSED	**Eating**	bulimia **(3)**				
		anorexia:hyporexia **(4)**				
		dysorexia (ranging from hyper to hypo) **(5)**				
		normal appetite **(1)**				
		bulimia with regurgitation and reingestion **(3)**				
	Drinking	eudipsia **(1)**				
		polydipsia (documented) **(5)**				
		chews at water without swallowing **(3)**				
		pushes the empty bowl around **(2)**				
	Somesthetic	normal grooming behaviour **(1)**				
		licking, chewing **(4)**				
		stereotypical chewing, turning **(5)**				
	Sleep	normal (or no change) **(1)**				
		increase, hypersomnia **(2)**				
		insomnia while sleeping **(3)**				
		wakes soon after going to sleep, unsettled upon waking **(5)**				
OUTWARD FOCUSED	**Exploratory behaviour**	normal **(1)**				
		simple inhibition **(2)**				
		increased with hypervigilance **(4)**				
		oral **(5)**				
		frequent avoidance behaviours **(3)**				
	Aggression	aggressiveness unchanged (with no relational problems) **(1)**				
		aggression in response to irritation **(3)**				
		aggression in response to fear **(4)**				
		aggression in response to fear and irritation **(5)**				
	Social skills	Steals, does not release stolen items **(5)**				
		bites without growling **(4)**				
		absence of submission **(2)**				
		lack of control when playing **(2)**				
		unchanged **(1)**				
	Specific skills	same response capacity (taking tiring into account) **(1)**				
		random responses **(3)**				
		more responses **(5)**				
	Physical examination	normal **(1)**				
		episodes of tachycardia/tachypnoea **(2)**				
		diarrhoea, colic **(2)**				
		dyspepsia **(2)**				
		increased urine output **(3)**				
		lick granuloma **(4)**				
		obesity **(4)**				
		PUPD **(4)**				
Total score						

Fig. 1. EDED questionnaire.

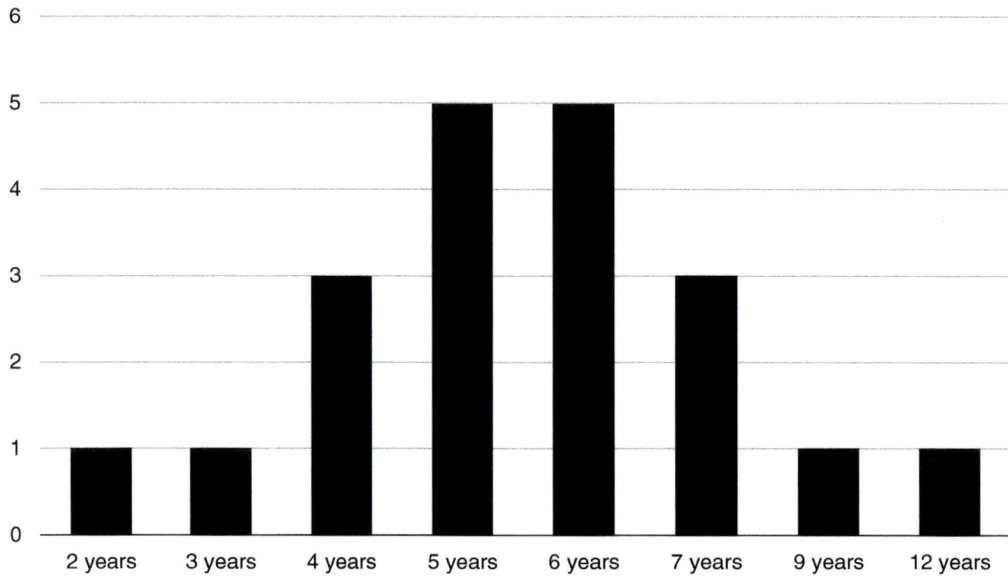

Fig. 2. Age distribution of the dogs with chronic gastric disorder.

Table 2. EDED results in ill dogs and their matched controls.

Name	Breed	Age	Sex	Sign	Score	Control
Jack	Airedale	5 years	male	dyspepsia	14	13
Hoppy	Griffon	9 years	male	vomiting	18	16
Sunny	WHWT	4 years	male	vomiting	19	12
Jazz	Yorkshire	7 years	male	vomiting	21	10
Minnie	Jack Russell	6 years	female	dyspepsia	30	9
Kenji	Labrador	6 years	male	vomiting	18	10
John	Poodle	7 years	male	dyspepsia	28	9
Guismo	Shih Tzu	5 years	male	vomiting	20	9
Olaf	Boxer	3 years	male	dyspepsia	23	14
Janie	Shih Tzu	6 years	female	vomiting	13	13
Newton	Shetland	4 years	male	vomiting	20	9
Nouki	WHWT	4 years	male	vomiting	18	10
Floyd	Labrador	5 years	male	vomiting	22	15
Gribouille	Yorkshire	12 years	male	vomiting	21	11
Lili	Yorkshire	7 years	female	vomiting	11	17
Noupy	Boxer	5 years	male	dyspepsia	21	12
Miss	Labrador	5 years	female	vomiting	20	9
Margot	Poodle	6 years	female	vomiting	26	12
Mozart	Poodle	6 years	male	vomiting	23	9
Moustique	Yorkshire	2 years	male	vomiting	24	11

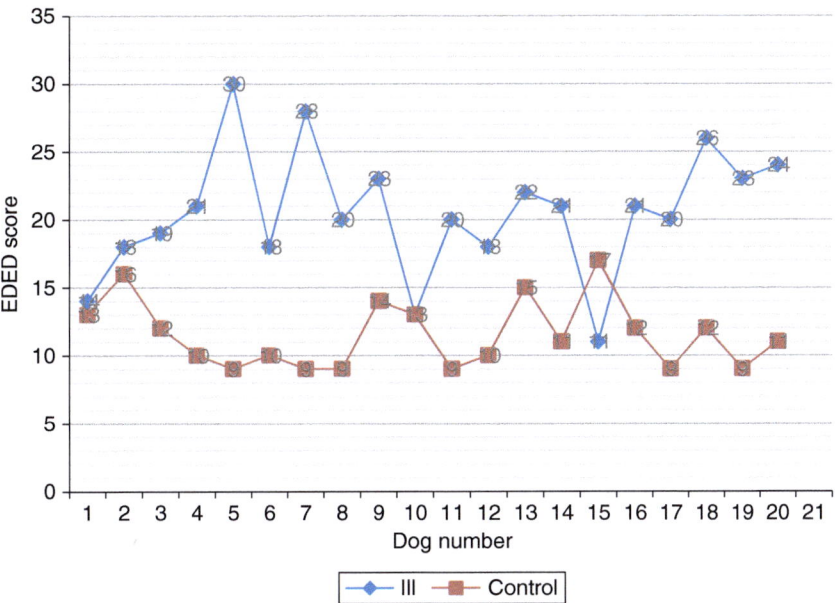

Fig. 3. Distribution of EDED scores of the ill and control dogs.

Anxiety can promote the proliferation of *Helicobacter pylori* in the stomach (Andreica-Sandica *et al.*, 2011; Cader *et al.*, 2015).

This difference is not seen in dogs. Even though *Helicobacter* spp. are present in dog stomachs, there is no significant difference in the rate of colonisation between dogs with gastrointestinal disease and healthy dogs (Lecoindre, 2001). Multiple species of *Helicobacter* are present in dogs, including *H. felis*, *H. bizzozeronii* and *H. salomoni*. They have not been shown to be pathogenic in dogs. Experimental infection is accompanied by gastritis and an immune response with no associated clinical signs (Lecoindre *et al.*, 1997).

The average age of the dogs in this study was 5.7 years. Chronic gastric disorders respond well to symptomatic treatment. These disorders do not increase the mortality risk for the animal. Considering the difficulty of motivating dog owners to perform a very comprehensive diagnostic, it is likely that the average age of the dogs in this study was higher than the average age of onset. Conversely, this type of examination requires a general anaesthetic; senior dog owners may prefer to treat them without a definitive diagnosis to avoid the risk of general anaesthesia. These factors could explain the Gaussian distribution of the study population (Fig. 2).

In our study of 20 dogs, 15 were males, and 5 were female. A study performed at the same time involved 489 dogs who were brought into six different veterinary surgeries for vaccination and reported that 55% of the dogs were male (intact and sterilised), whereas 45% were female (intact and sterilised) (Beaumont, 2002). The sex ratio was the opposite for dogs presenting with CILBD, with females representing more than 56% of the affected dogs (Reiwald *et al.*, 2013). Our results suggest that males are overrepresented in our 'ill' population; however, a larger study is needed to confirm this finding.

Small breeds were better represented than large and medium-sized breeds. Smaller dogs often live in closer contact with their owners than do large dogs. It is possible that vomiting or dyspeptic episodes go unnoticed if the dog lives primarily outdoors. Yorkshire Terriers, Poodles and Labrador Retrievers were the breeds that were most represented in our study (45%, 9/20). However, when this distribution is compared to the distribution of breeds within the French canine population, it appears that these were the three most common breeds in France at the time of the study (distribution based on a study by Facco/Sofres, unpublished results).

The statistically significant difference ($p<0.05$) between the average EDED scores of the two populations allows us to conclude that anxiety is a possible aetiology of chronic gastric diseases.

The pathophysiology of chronic gastric disorders resulting from anxiety remains largely unclear. However, experimental work has demonstrated a link between acoustic stress and changes in gastrointestinal motility in dogs (Gué, 1989). Dogs were fitted with headphones that played music with a high sound intensity. This sound induced a stress reaction, as demonstrated by an accelerated heart rate and an increase in plasma cortisol levels. In fasting dogs, experiencing acoustic stress for 1 hour results in inhibition of gastric motility (though intestinal motility was not affected). During the postprandial phase, applying acoustic stress for 2 hours also causes gastric disturbances, as demonstrated by the delayed gastric emptying of the solid phase of a meal, associated with an increase in plasma levels of digestive hormones (gastrin, somatostatin and pancreatic polypeptide). The stimulation of the secretion of digestive hormones is prolonged beyond the duration of the stressful event. This prolongation is likely to be mediated not only by the activation of the hypothalamic–pituitary–adrenal (HPA) axis but other pathways as well. Prolonged postprandial gastric and intestinal motility, leading to both delayed gastric evacuation and hypersecretion of digestive hormones, has been observed in dogs.

The neurotransmitters and nerve pathways involved in the pathogenesis of gastric disorders linked to stress have been the subject of a previous study (Gué, 1989). The pneumogastric nerves may be involved in motility disorders. Also, corticotropin-releasing hormone (CRH) could help initiate or mediate gastrointestinal perturbations induced by stress. Corticotropin-releasing hormone could act through the supraspinal structures that regulated gastrointestinal motility, and not through the HPA axis. However, this work did not focus on digestive disorders linked to acute stress. Chronic gastric disorders arising from anxiety could occur due to other pathophysiologic mechanisms. Multiple studies conducted in rats and mice seem to indicate that chronic stress could lead to intestinal inflammation, associated with the mobilisation of mast cells and accumulated secretion of CRH and acetylcholine. This inflammation could induce an increase in intestinal permeability (Qiu *et al.*, 1999; Velin *et al.*, 2004; Glaser and Kiekolt-Glaser, 2005).

If anxiety can cause chronic gastric disease, it is also possible that the causality could be reversed in some circumstances. Female rats are prone to developing signs of anxiety after iatrogenic gastritis (Luo *et al.*, 2013). In addition, some behavioural problems, such as excessive licking of surfaces, are associated with gastrointestinal disorders in dogs, and not related to anxiety (Bécuwe-Bonnet *et al.*, 2012).

There was no significant difference between the EDED scores of dogs with dyspepsia and dogs exhibiting vomiting. This supports the homogeneity of the ill dog group based on the criteria investigated. Vomiting and dyspepsia represent two potential clinical manifestations of canine anxiety. Behavioural problems can, therefore, be included in the differential diagnosis of chronic gastric disorders when accompanied by vomiting or dyspepsia.

Conclusion

The differential diagnosis of chronic gastric disease should include anxiety, and not only as an exclusion diagnosis. Scoring chronic and relapsing dogs on an EDED scale can save time. Treating anxiety improves the outcome of these dogs. Vomiting and dyspepsia are clinical signs indicating anxiety as a behavioural pathology. These results should be corroborated in larger studies to confirm that a causal link exists between anxiety and chronic gastric disease in dogs.

References

Addolorato, G., Mirijello, A., D'Angelo, C., Leggio, L., Ferrulli, A., Abenavoli, L., Vonghia, L., Cardone, S., Leso, V., Cossari, A., Capristo, E. and Gasbarrini, G. (2008) State and trait anxiety and depression in patients affected by gastrointestinal diseases: psychometric evaluation of 1641 patients referred to an internal medicine outpatient setting. *International Journal of Clinical Practice* 62, 1063–1069.

Andreica-Sandica, B., Panaete, A., Pascanu, R., Sarban, C. and Andreica, V. (2011) The association between Helicobacter Pylori chronic gastritis, psychological trauma and somatization disorder. a case report. *Journal of Gastrointestinal and Liver Diseases* 20, 311–313.

Beaumont, G. (2002) La Virgule - étude des représentations. Memory supported in the context of the inter-school diploma of behavioral veterinarians. Available at: http://www.zoopsy.com/prive/memoires/ (accessed 30 May 2017).

Bécuwe-Bonnet, V., Bélanger, M.C., Frank, D., Parent, J. and Hélie, P. (2012) Gastrointestinal disorders in dogs with excessive licking of surfaces. *Journal of Veterinary Behavior: Clinical Applications and Research* 7, 194–204.

Cader, J., Domagala, Z., Paradowski, L., Rymaszewska, J., Blonski, W. and Sajewicz, Z. (2015) Is there any relation of Helicobacter pylori infection to anxiety and depressive symptoms? *Polish Gastroenterology* 14, 397–401.

Elwood, C., Devauchelle, P., Elliott, J., Freiche, V., German, A.J., Gualtieri, M., Hall, E., Den Hertog, E., Neiger, R., Peeters, D., Roura, X. and Savary-Bataille, K. (2010) Emesis in dogs: a review. *Journal of Small Animal Practice* 51, 4–22.

Glaser, R. and Kiekolt-Glaser, J. (2005) Stress-induced immune dysfunction: implications for health. *Nature Reviews Immunology* 5, 243–251.

Goodwin, R.D., Cowles, R.A., Galea, S. and Jacobi, F. (2013) Gastritis and mental disorders. *Journal of Psychiatric Research* 47, 128–132.

Gué, M. (1989) Nature et origine des troubles moteurs gastrointestinaux liés aux stress acoustique et thermique: rôle de la corticolibérine. Doctoral dissertation. *Toulouse 3*.

Heams, E. (1996) Maladie ulcéreuse et gastrites à l'heure d'Helicobacter pylori. *Thérapie* 51, 309–318.

Lecoindre, P. (2001) Les maladies de l'estomac. *Pratique Médicale et Chirurgicale de l'Animal de Compagnie* 36, 351–360.

Lecoindre, P. and Gaschen, F.P. (2011) Chronic idiopathic large bowel diarrhea in the dog. *Veterinary Clinics of North America: Small Animal Practice* 41, 447–456.

Lecoindre, P., Chevalier, M., Peyrol, S., Boude, M., Labigne, A., Lamouliatte, H. and Pilet, C. (1997) Helicobacter infections of man and of domestic carnivores: comparative data. *Bulletin de L'Académie Nationale de Médecine* 181, 431–439.

Luo, J., Wang, T., Liang, S., Hu, X., Li, W. and Jin, F. (2013) Experimental gastritis leads to anxiety- and depression-like behaviors in female but not male rats. *Behavioral and Brain Functions* 9, 46.

Overall, K.L., Dunham, A.E. and Frank, D. (2001) Frequency of nonspecific clinical signs in dogs with separation anxiety, thunderstorm phobia, and noise phobia, alone or in combination. *Journal of the American Veterinary Medical Association* 4, 467–473.

Pageat, P. (1990) Sémiologie en pathologie comportementale canine 1ère et 2ème parties. *Point Veterinary* 22, 128–129.

Qiu, B.S., Vallance, B.A., Blennerhasset, P.A. and Collins, S.M. (1999) The role of CD4 lymphocytes in the susceptibility of mice to stress-induced reactivation of experimental colitis. *Nature Medicine* 5, 1178–1182.

Reiwald, D., Pillonel, C., Villars, A.M. and Cadore, J.L. (2013) Anxiété et entéropathies inflammatoires chroniques idiopathiques chez le chien. *Revista De Medicina Veterinaria* 164, 145–149.

Tiira, K., Sulkama, S. and Lohi, H. (2016) Prevalence, comorbidity, and behavioral variation in canine anxiety. *Journal of Veterinary Behavior* 16, 36–44.

Velin, A.K., Ericson, A.C., Braaf, Y., Wallon, C. and Soederholm, J.D. (2004) Increased antigen and bacterial uptake in follicle associated epithelium induced by chronic psychological stress in rats. *Gut* 53, 494–500.

Interaction of Health and Behaviour Problems in Dogs

Maya Braem Dube[1]*, Lucy Asher[2], Hanno Würbel[1] and Luca Melotti[1]

[1]Division of Animal Welfare, Veterinary Public Health Institute, Vetsuisse Faculty, University of Berne, Berne, Switzerland; [2]Centre for Behaviour and Evolution, Institute of Neuroscience, Newcastle University, Newcastle, UK

Funding: The study was funded by the Margaret and Frances Fleitmann Foundation.

Conflict of interest: The authors declare no conflict of interest.

Keywords: dog, behaviour problem, health

Introduction

The interaction between health and behaviour problems plays an important role in veterinary behaviour medicine. Physical problems, such as pain (Barcelos *et al.*, 2015) or gastrointestinal problems (Bécuwe-Bonnet *et al.*, 2012) have been shown to be linked to behaviour problems. The aim of this study was to investigate this association further.

Material and Methods

Using an online survey, information was collected on dog and owner demographics, the dog's health and the occurrence of behaviour problems. The relationship between the presence of behaviour and health problems was assessed using a chi-squared test.

Results

In total, 3646 completed questionnaires were analysed. Fifty per cent of the dogs were male (67% neutered) and 50% female (76% neutered). Owners of 16.5%

---

* Corresponding author: maya.braem@vetsuisse.unibe.ch

of the dogs indicated that their dogs had or had had a physical problem (18.8% musculoskeletal, 14% neuro-endocrino-immunological, 8.7% gastro-intestinal, 7.2% dermal, 42.2% two or more, 9.2% other physical problems). Behaviour problems were reported in 41.4% of the dogs (19.3% aggression, 30.6% fear, 10.5% aggression and fear, 8.4% excessive, 29.5% other behaviour problems). A significant correlation was found between showing a behaviour problem and having a health issue (χ^2 (1, N = 3646) = 37.33, p < 0.001). Relationships between specific types of behaviour and physical problems are descriptively discussed.

Conclusion

Physical problems were correlated with an increased likelihood of behaviour problems in dogs. Although these data do not allow for deduction of direct causality they give a clear indication that health problems should be considered when diagnosing behaviour problems in dogs.

References

Barcelos, A.M., Mills, D.S. and Zulch, H. (2015) Clinical indicators of occult musculoskeletal pain in aggressive dogs. *Veterinary Record* 176, 10.1136/vr.102823.

Bécuwe-Bonnet, V., Bélanger, M.C., Frank, D., Parent, J. and Hélie, P. (2012) Gastrointestinal disorders in dogs with excessive licking of surfaces. *Journal of Veterinary Behavior* 7, 194–204.

Keynote Presentation: A Multimodal Approach to Resolving Tension Between Cats in the Same Household: A Practical Approach

Sᴀʀᴀʜ E. Hᴇᴀᴛʜ*

Behavioural Referrals Veterinary Practice, Upton Chester, UK

Conflict of interest: The author declares no conflict of interest.

Keywords: behaviour, cats, inter-cat conflict, stress

Introduction

Cats are increasing in popularity in many Western countries, and multi-cat households are very common. While many of these households are successful and harmonious, it is recognised that some owners encounter difficulty, either regarding the presentation of behavioural responses which cause them concern or regarding the physical diseases seen in one or more of the cats. Common physical disease presentations from multi-cat environments include idiopathic cystitis (Kruger *et al.*, 2009) and infectious disease (Speakman, 2005). Also, obesity has been shown to be influenced by feedings styles (German and Heath, 2016) which in turn may be affected by the presence of more than one cat in the household. Reported problematic behaviour presentations include urine spraying (Carney *et al.*, 2014) and inter-cat conflict (Pachel, 2014).

Manifestations of Feline Tension

For many owners, an appreciation of tension between cats in the household is dependent on the presence of actively confrontational behaviours such as hissing,

---

* heath@brvp.co.uk

swiping, batting, chasing and biting. The use of the term inter-cat aggression implies the existence of physical conflict and owner reporting of problems of social interaction between their cats has been shown to be dependent on the presence of behaviours which they identify as 'aggression' (Levine *et al.*, 2005). Feline tension is not always manifested through active behaviours. Feline social behaviour influences the motivation for cats to avoid out and out physical confrontation and, in situations where cats are socially incompatible, other more passive behaviours are often indicative of tension between them. Distance increasing behavioural interactions, such as staring or actively avoiding, are commonly overlooked while interactions which appear to exclude one or more cats from certain areas of the house are misinterpreted as signs of dominance and bullying (Pachel, 2014).

Clinical Relevance

As the number of multi-cat households increases, the welfare implications of unrecognised social incompatibility between cats also increases. Owners largely select members of a feline household based on their desires and preferences. Cats are solitary survivors who restrict their social interactions to members of their social group (Bradshaw *et al.*, 2012a) and even then, do not share certain behaviours, such as eating. Survival is the overwhelming motivation and, since social interaction is not an integral part of that process, cats can take and leave a social contact in a way that humans find difficult to understand. A lack of understanding of feline social behaviour and tendency for owners to attribute their perceptions to their pets can lead to the establishment of households where the cats are living in a state of chronic social tension. Optimising the environment in ways that respect feline social behaviour is the key to reducing the negative impact of social tension and increasing the possibility of harmony in multi-cat households. This harmony may not be entirely in keeping with the owner's expectations and education about natural feline behaviour will be needed to minimise owner frustration. In many cases, the owners would like to see feline friendships develop between the cats within the household but in most cases, this is unrealistic. Only the cats themselves can determine the feline social relationships, and when humans bring cats under the same roof who have no biological reason to be friends, they must accept that the aim is feline tolerance rather than friendship. The goal is for all of the cats living at the same address to be able to move freely around the household territory and to access essential resources without the risk of passive or active hostility. Achieving this aim will significantly improve the welfare of the cats but will also have a positive effect on the quality of life of the owners.

Optimising the Feline Environment

The basis of a multimodal approach to resolving feline tension is the meeting of feline environmental needs within the domestic environment (Ellis *et al.*, 2013). The American Association of Feline Practitioners (AAFP) and the International Society of Feline Medicine (ISFM) Feline Environmental Needs

Guidelines suggest a concept of five pillars, which provide the framework for an optimal feline environment.

Pillar 1. Provide a safe place.
Pillar 2. Provide multiple and separated key environmental resources.
Pillar 3. Provide opportunity for play and predatory behaviour.
Pillar 4. Provide positive, consistent and predictable human–cat social interaction.
Pillar 5. Provide an environment that respects the importance of the cat's sense of smell.

Understanding social groups

These five pillars provide a useful starting point when veterinary surgeons in general practice are faced with a client reporting behavioural or physical health issues in a multi-cat environment. To establish a practical programme of change for the owner the first step is to understand the social groups which exist within the household. Many owners are unaware of the relationships between their cats, and it is very common to report that the cats are friends or even 'love each other' on the basis that there has never been any overt hostility shown between them. 'Friendships' are indicated by the presence of affiliative behaviours rather than the absence of confrontation. Affiliative behaviours in the feline context are allorubbing and allogrooming. To have a better understanding of how to proceed in a problematic multi-cat household, or to optimise any home with more than one feline resident, it is useful for owners to create an 'affiliative behaviour map'. This is achieved by closely observing interactions between the cats over a seven-day period and recording any interactions of allorubbing or allogrooming. This information can be visually stored by writing the names of the cats on a piece of paper and drawing arrows from the initiator to the receiver of any rubbing or grooming interactions. It can be helpful for these arrows to be in different colours, to make the reading of the map easier. It will also be beneficial to place every arrow related to the same interaction in the same direction between two individuals next to each other as this will result in the breadth of the arrow representing the frequency of those interactions (Fig. 1).

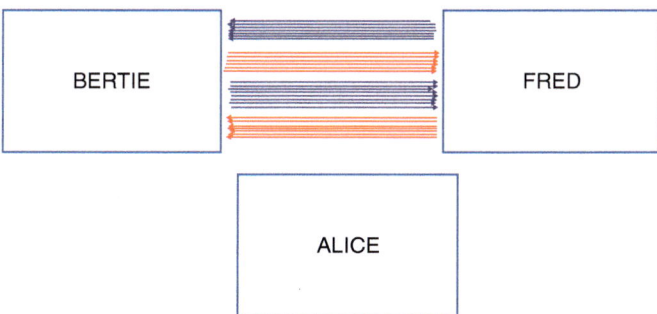

Fig. 1. Illustration of inter-cat relationship in the same household.

During the process of creating the 'affiliative behaviour map' owners will often become more alert to their cats' behaviour, and it can be helpful to ask them also to record any hostile behaviours that they witness. These may take the form of active interactions, such as hissing, swiping or biting, but should also include passive interactions of staring and distancing. This exercise can help owners to become more aware of the interactions that take place between their cats, but also to improve their understanding of the level of tension between them.

Practical implications of the five pillars

It is important to adopt a multimodal approach to clinical cases of feline household tension in order to meet the requirements of the five environmental needs pillars. The provision of a safe place refers to the establishment of a core territory for each of the social groupings of cats in the household. Since the core territory is the central, or most important, part of the territory regarding survival, the provision of multiple and separated key environmental resources is key to establishing it. Resources for domestic cats include the obvious ones of food, water, resting places and latrines but also include an ability to traverse the territory safely and to engage in coping strategy behaviours, such as hiding and elevation. Studies have shown that expansion and increased availability of three-dimensional space can significantly improve feline welfare (Desforges et al., 2016) and this can be a relatively straightforward practical approach in multi-cat households. Increasing availability of safe and secure resting places and sufficient, appropriately positioned resource stations for food, water and latrines has also been suggested (Rochlitz, 2009; Amat et al., 2016).

Cats are predatory mammals, and the ability to engage in predatory behaviour both in the context of food acquisition and in play is an important part of optimising the feline environment (Bradshaw et al., 2012b).

The provision of social company is not a survival resource for cats in the same way as it is for a socially obligated mammal, such as a dog or human. However, in the domestic context cats are socialised with humans from an early age and will often incorporate human animals within their social groups (Bradshaw, 2013). They will engage in affiliative behaviours toward them, and the provision of social contact in a consistent and predictable fashion is therefore necessary. Differences in human and feline expectations in the context of social relationships need to be considered (Downey and Ellis, 2008).

The final pillar refers to the need to respect the importance of the cat's sense of smell. Cats use olfactory communication to define their territories and to express emotional information. Commercial products are available which aim to mimic natural feline olfactory signals in both of these contexts. One fraction of the feline facial pheromone complex, F3, has been advocated as a means of increasing the perception of a safe territory and thereby reducing the incidence of urine spraying in a domestic context (Ogata and Yukari, 2001). It has also been explored in the context of managing cases of feline idiopathic cystitis (Gunn-Moore and Cameron, 2004). The use of F3 in situations of multi-cat household tension has also been reported. The use of the appeasing pheromones in other species such as dogs, pigs and horses has been documented for some time, but recently feline appeasing phero-mone has become commercially available. The effect of this olfactory signal on signs of social tension in cats has been explored (Cozzi *et al.*, 2010) and the commercial product is recommended as a means of improving feline harmony (De Porter *et al.*, 2014). As with any approach to reducing social tension in a multi-cat household, it is important to respect natural feline social behaviour and to guard against unrealistic expectations of owners in terms of encouraging friendships to develop between socially incompatible cats. Optimising the environment in terms of respecting the role of scent communication in cats can help to increase the personal safety of each cat within the household and, in conjunction with other environmental adjustments, the aim is to increase tolerance and reduce tension.

Conclusions

Feline tension in multi-cat households is an important welfare issue. It has both behavioural and physiological consequences and is, therefore, an excellent example of the importance of behavioural medicine and animal well-being in the veterinary context. The five pillars of feline environmental needs provide a useful framework for understanding the challenges that many domestic cats face when living in multi-cat households. The application of practical strategies to optimise the domestic environment can reduce behavioural signs of tension between cats in the same household, and improve not only their welfare but also the well-being of their owners.

References

Amat, M., Camps, T. and Manteca, X. (2016) Stress in owned cats: behavioural changes and welfare implications. *Journal of Feline Medicine & Surgery* 18(8), 577–586.

Bradshaw, J. (2013) More than a feline. *New Scientist* 219(2934), 44–47.

Bradshaw, J.W.S., Brown, S.L. and Casey, R. (2012a) Social behavior. In: *The Behaviour of the Domestic Cat*. CAB International, Wallingford, UK, pp. 142–160.

Bradshaw, J.W.S., Brown, S.L. and Casey, R. (2012b) Hunting and predation. In: *The Behaviour of the Domestic Cat*. CAB International, Wallingford, UK, pp. 128–141.

Carney, H.C., Sadek, T.P., Curtis, T.M., Halls, V., Heath, S.E., Hutchinson, P., Mundschenk, K. and Westropp, J.L. (2014) AAFP and ISFM guidelines for diagnosing and solving house-soiling behavior in cats. *Journal of Feline Medicine & Surgery* 16, 579–598.

Cozzi, A., Monneret, P., Lafont-Lecuelle, C., Bougrat, L., Gaultier, E. and Pageat, P. (2010) The Maternal cat appeasing pheromone: exploratory study of the effects on aggressive and affiliative interactions in cats. *Journal of Veterinary Behavior* 5(1), 37–38.

De Porter, T.L., Lopezb, A. and Ollivierc, E. (2014) Evaluation of the efficacy of a new pheromone product versus placebo in the management of feline aggression in multi-cat households. *Proceedings Veterinary Behavior Symposium* 17–18.

Desforges, E.J., Moesta, A. and Farnworth, M.J. (2016) Effect of a shelf-furnished screen on space utilisation and social behaviour of indoor group-housed cats (*Felis silvestris catus*). *Applied Animal Behaviour Science* 178, 60–68.

Downey, H. and Ellis, S. (2008) Tails of animal attraction: Incorporating the feline into the family. *Journal of Business Research* 61(5), 434–441.

Ellis, S.L., Rodan, I., Hazel, C., Carney, H.C., Heath, S.E., Rochlitz, I., Shearburn, L.D., Sundahl, E. and Westropp, J.L. (2013) AAFP and ISFM Feline Environmental Needs Guidelines. *Journal of Feline Medicine & Surgery* 5, 219–230.

German, A.J. and Heath, S.E. (2016) Feline obesity – a medical disease with behavioural influences. In: Rodan, I. and Heath, S.E. (eds) *Feline Behavioural Health and Welfare*. Elsevier, St Louis, Missouri, pp. 148–161.

Gunn-Moore, D. and Cameron, M.E. (2004) A pilot study using synthetic feline facial pheromone for the management of feline idiopathic cystitis. *Journal of Feline Medicine & Surgery* 6(3), 133–138.

Kruger, J.M., Osborne, C.A. and Lulich, J.P. (2009) Changing paradigms of feline idiopathic cystitis. *Veterinary Clinics: Small Animal Practice* 39(1), 15–40.

Levine, E., Perry, P., Scarlett, J. and Houpt, K.A. (2005) Intercat aggression in households following the introduction of a new cat. *Applied Animal Behaviour Science* 90(3–4), 325–336.

Ogata, N. and Yukari, T. (2001) Clinical trial of a feline pheromone analogue for feline urine marking. *Journal of Veterinary Medical Science* 63(2), 157–161.

Pachel, C.L. (2014) Intercat aggression: restoring harmony in the home: a guide for practitioners. *Veterinary Clinics: Small Animal Practice* 44(3), 565–579.

Rochlitz, I. (2009) Basic requirements for good behavioural health and welfare of cats. In: Horwitz, D.F. and Mills, D. (eds) *BSAVA Manual of Canine and Feline Behavioural Medicine*. Brit Small Anim Vet Assoc, Gloucester, UK, pp. 35–48.

Speakman, A. (2005) Management of infectious disease in the multi-cat environment. *In Practice* 27(9), 446–453.

Effect of a Synthetic Feline Pheromone for Managing Unwanted Scratching in Domestic Cats

Valarie V. Tynes[1]*, Alexandra Beck[2], Xavier De Jaeger[2] and Jean-Francois Collin[2]

1Ceva Animal Health, Lenexa,Kansas, USA; 2Ceva Santé Animale, Libourne, France

Conflict of interest: All authors are employed by Ceva Animal Health who funded this research.

Keywords: behaviour, cat, claw marking, feline interdigital semiochemical, pheromone, scratching

Introduction

Scratching of objects in the environment is a normal behaviour for domestic cats but is often problematic for pet owners. Even in the presence of a scratching post, many cats will continue to scratch household furnishings (Landsberg, 1991; Wilson *et al.*, 2016). When scratching a surface, cats produce a visible mark as well as a chemical message (a semiochemical released from their interdigital area). The present study tested a solution containing a synthetic analogue of the feline interdigital semiochemical (F.I.S.)[1] to determine if it could effectively redirect cat scratching behaviour towards a scratching post. The coloured product when applied on a post actually mimics both the chemical (F.I.S.) and visual (lacerations) cues naturally left by scratch marks, to encourage cats to scratch again on the post. Authors hypothesised that the application of the product to a scratching post would stimulate the use of the post while concomitantly limiting or even stopping scratching on undesired surfaces.

---

* Corresponding author: valarie.tynes@ceva.com

Materials and Methods

A total of 166 cats from 117 households that had demonstrated scratching on vertical surfaces within the past year were included. Twenty-nine 'newly adopted' adult cats or kittens that had demonstrated unwanted scratching were also included. Each cat served as their own control. All owners were given a new scratching post (pole type scratching post covered with rope) and instructed to place it either near a frequently used area for scratching, or close to a cat's sleeping or relaxing area. Owners were instructed to remove all other posts from the home. The pheromone product was supplied in single dose pipettes to be applied by drawing a single vertical line on the scratching post once daily on Days 0–6, 14, 21 and 28. Owners were instructed to complete daily questionnaires and their data were collected during weekly phone interviews. The scratching frequency was analysed statistically using a mixed model with a time effect as fixed covariate. The proportion of cats who stopped scratching used a logistic model with a time effect as fixed covariate and a random animal effect with a logit link function (p = probability of not scratching). The ANOVA and Student t-tests were derived from those models.

Results

The number of cats who stopped scratching on vertical and horizontal surfaces significantly increased between baseline and D7. After 28 days, a further significant improvement was observed, with 74% of the cats with established unwanted scratching behaviour having completely stopped scratching on vertical surfaces other than the provided scratching post. A significant decrease ($p < 0.001$) in scratching on horizontal surfaces was also observed. These effects were similar for cats with a scratching post prior to the study and those who did not have a scratching post, allowing for the separation of the effects of the tested solution from the new post introduction. 'Newly adopted' cats also demonstrated a significant decrease ($p < 0.001$) in scratching on vertical surfaces in the home by D7 and even D28 of the study. Cats were observed for two more weeks after treatment application was stopped, and no relapse was observed. While decreasing scratch marks on household surfaces, cats also showed a clear interest in their new scratching post treated with F.I.S. product: after only 7 days of treatment, 79% of the 'scratching cats' and 87% of the 'newly adopted' cats had used it for scratching.

Conclusion

The application of this interdigital pheromone analogue solution appears to be an innovative and effective solution to the scratching problem in cats. Moreover, this innovative treatment solution presents a natural and humane alternative to other current options that is respectful towards the cat's ethology.

Note

[1] Marketed as Feliscratch by FELIWAY® (Ceva Santé Animale, France).

References

Landsberg, G.M. (1991) Feline scratching and destruction and the effects of declawing. *Veterinary Clinics of North America: Small Animal Practice* 21, 265–279.

Wilson, C., Bain, M., DePorter, T., Beck, A., Grassi, V. and Landsberg, G. (2016) Owner observations regarding cat scratching behavior: an internet-based survey. *Journal of Feline Medicine & Surgery* 18, 791–797.

Relationship Among Cat–Owner Bond, Cat Behaviour Problems and Cat Environment Conditions: A Study with 1553 Spanish Cat Owners

Natalia Bulgakova[1]*, Sandra Burgos[1], Paula Calvo[1], Jonathan Bowen[1,2] and Jaume Fatjó[1]

[1]Chair Affinity Foundation Animals and Health, Universitat Autònoma de Barcelona, Spain; [2]Royal Veterinary College, University of London, UK

Conflict of interest: The authors declare no conflict of interest.

Keywords: cat–owner bond, cat behaviour problems, cat welfare

Introduction

Owners' bonds with their cats may affect awareness of cats' well-being and adherence to behavioural treatment. Our aim was to explore relationships between the cat–owner bond, cat welfare conditions and cat behaviour problems.

Materials and Methods

We conducted an online survey, which included the Cat Owner Relationship Scale[1] (CORS), a cat welfare scale, and a cat behaviour problem assessment. Cat owners were recruited through social media networks. We analysed correlations between scores for CORS, welfare and behaviour problems. We also analysed all 3 dimensions of CORS (interaction, perceived costs and emotional bond) in relation to the presence of behaviour problems (Mann–Whitney U Test).

* Corresponding author: natalia@holistic-cat.com

Results

We obtained 1553 completed questionnaires. CORS was directly correlated with welfare results. In relation to behaviour problems, CORS was correlated with aggression towards family members: inversely correlated with the frequency and intensity of the problem, and was directly correlated with owners' tolerance of the problem. When comparing CORS results between owners of aggressive and non-aggressive cats, perceived costs of ownership were significantly higher for those with aggressive cats.

Conclusion

In conclusion, a higher cat–owner bond seems to be directly related to better cat welfare and higher tolerance of cat behaviour. Some behaviour problems appear to have a negative effect on the perceived costs of owning a cat, which could lead to relinquishment. Hence, measuring the cat–owner relationship could be an interesting tool to consider in behavioural medicine.

Note

[1]Howell, T., Bowen, J., Fatjó, J., Calvo, P., Holloway, A. and Bennett, P. Development of the cat-owner relationship scale (CORS). Unpublished data.

1 cat per household. 663 participants

Neuter status		Behaviour problems		
Neutered	Not neutered	No problems	Problems	
88%	12%	71%	Total problems	29%
			Soiling problems	8%
			Inter-cat aggression	0%
			Human directed aggression	16%
			2 or 3 problems together	5%

Demographic data about participants for 1 cat household

Sex		Age					Level of education	
Man	Woman	Man		Woman				
		18–35 y.o.	36–45 y.o.	18–35 y.o.	36–45 y.o.	Household without children	University Degree	Certificate of Higher Education (HNC)
11%	89%	53%	33%	53%	30%	87%	60%	17%

2 cats per household. 535 participants

Neuter status		Behaviour problems		
Neutered	Not neutered	No problems	Problems	
94%	6%	62%	Total problems	38%
			Soiling problems	11%
			Inter-cat aggression	11%
			Human directed aggression	3%
			2 or 3 problems together	13%

Demographic data about participants for 2 cats household

Sex		Age					Level of education	
Man	Woman	Man		Woman				
						Household without children	University Degree	Certificate of Higher Education (HNC)
		18–35 y.o.	36–45 y.o.	18–35 y.o.	36–45 y.o.			
14%	86%	43%	37%	54%	28%	87%	55%	20%

3 cats per household. 171 participants

Neuter status		Behaviour problems		
Neutered	Not neutered	No problems	Problems	
93%	6%	58%	Total problems	42%
			Soiling problems	11%
			Inter-cat aggression	19%
			Human directed aggression	0%
			2 or 3 problems together	12%

Demographic data about participants for 3 cats household

Sex		Age					Level of education	
Man	Woman	Man		Woman				
						Household without children	University Degree	Certificate of Higher Education (HNC)
		18–35 y.o.	36–45 y.o.	18–35 y.o.	36–45 y.o.			
13%	87%	36%	41%	44%	34%	83%	51%	18%

4 cats per household. 86 participants

Neuter status		Behaviour problems		
Neutered	Not neutered	No problems	Problems	
92%	7%	61%	Total problems	39%
			Soiling problems	15%
			Inter-cat aggression	17%
			Human directed aggression	0%
			2 or 3 problems together	7%

Demographic data about participants for 4 cats household

Sex		Age					Level of education	
Man	Woman	Man		Woman				
						Household without children	University Degree	Certificate of Vocational Education (VEC)
		18–35 y.o.	36–45 y.o.	18–35 y.o.	36–45 y.o.			
5%	95%	50%	50%	43%	37%	90%	48%	18%

5 and 6 cats per household. 58 participants

Neuter status		Behaviour problems		
Neutered	Not neutered	No problems	Problems	
93%	7%	52%	Total problems	48%
			Soiling problems	14%
			Inter-cat aggression	27%
			Human directed aggression	0%
			2 or 3 problems together	7%

Demographic data about participants for 5 and 6 cats household

Sex		Age					Level of education	
Man	Woman	Man		Woman				
						Household without children	University Degree	Certificate of Vocational Education (VEC)
		18–35 y.o.	36–45 y.o.	18–35 y.o.	36–45 y.o.			
3%	97%	50%	50%	34%	29%	95%	53%	16%

Keynote Presentation:
The Importance of the Welfare of Research Animals to Maximise the Quality of Behavioural Research: Do We Measure True Behaviours?

PATRICK PAGEAT*

IRSEA and E.I. Purpan, Quartier Salignan, France

Conflict of interest: The author declares no conflict of interest.

Keywords: welfare, ethology, research, development, socialisation

Despite the remarkable development of ethology, welfare science and behavioural medicine, our understanding of many behaviours is still limited. This lack of knowledge is much deeper when we try to discuss underlying mechanisms, development and functionality. The access to such information requires studying the target species in controlled conditions, which do not represent the actual environment of pet, farm or wild species. Moreover, the versatility of behaviours, as well as the inter-individual variability, lead the researchers to develop protocols that associate physiological and behavioural parameters. The resulting risk is that researchers describe behaviours and physiological variation that are based on the unnatural environment to that species. Additionally, it is possible that these environments may limit the animal and reduce its welfare, thus affecting our knowledge and results of the study. The purpose of this lecture is to discuss possible ethical strategies to prevent or limit such bias.

The three 'R's (Replacement, Reduction and Refinement) define an attractive framework for the organisation of animal research. In behavioural studies, Replacement is almost impossible. Reduction poses a difficulty when focusing on the number of individuals. In eusocial species, the social group should be regarded as a fundamental need, and thus counted as the experimental unit instead of

* p.pageat@group-irsea.com

individuals. Refinement is the key concept, which could be understood and accepted, even by the most reluctant researcher. A true description of behaviours requires that animals be placed not only in an appropriate environment but also their socialisation (intraspecific and interspecific for domestic species) and their social life address the unique needs of that species. Studies often evaluate stress-related reactions, not only natural behaviours. These studies lead researchers to keep their subjects in controlled environments, which may mean isolated animals in poor environments. Not only does this pose an ethical concern, but it also leads to the development of chronic stress-related responses (behavioural, cognitive and physical) that affect the quality and validity of the study results. Results of such study cannot be effectively generalised and widely applied, given the adverse effect the study had on the subjects.

Therefore, controlling the quality of the study protocol and design is paramount. There are several factors the researcher must consider when designing a scientifically sound study that yields valid results. When studying a group of animals, the social structure is important to remember; adding more individuals to an existing group may change the social organisation and does not necessarily create a functional social group. Attention should be given to the physical environment in which the animals live, and the study takes place, and effort must be given to optimise it. The type, and number, of experimental manipulations during the study must be considered as well.

Ultimately, research animals are under stressful conditions including different environments, manipulation (e.g. blood sampling, cardiac monitors, EEG electrodes and handling), cognitive testing, unfamiliar people and their behaviours, and noise. One possible solution to this problem is preparing animals, and habituating them to experimental conditioning. For example, before research, the scientists can train the animals to perform a basic behaviour that may be required later during the study (e.g. sit calmly for venepuncture or fitting with cardiac monitors). Moreover, habituating the animals to the personnel who will participate in the study is essential. Ideally, research institutes would have the workforce to care for these animals throughout their stay at the facility and prepare the animal year-round, regardless of the time and duration of the research. This approach has two fundamental benefits. First, the people in charge of the animals have the control over the experiment and the life quality of the animals (i.e. continuous Refinement). Second, the likelihood of euthanising the animal following the study will be minimised, as these animals are highly trained and valuable (Reduction). Ultimately, maintaining high-quality care, matching the animal's needs, and reducing anxiety and fear, leads to better study results with high validity.

Keynote Presentation: Making Animal Welfare Sustainable – Human Behaviour Change for Animal Behaviour: The Human Element

Jo White[1,2]* and Suzanne Rogers[2]

[1]Progressive Ideas, www.progressiveideas.co.uk; [2]Human Behaviour Change for Animals, www.hbcanimalwelfare.com

Conflict of interest: The authors declare no conflict of interest.

Keywords: human behaviour change, animal behaviour, animal welfare

Those working in veterinary behaviour medicine and animal welfare continue to deliver groundbreaking work that provides a greater understanding of the possible reasons why animals behave as they do, together with insights into human–animal relationships and animals welfare needs. However, the challenge of ensuring that these important findings are delivered at the coal face by those interacting and impacting upon animals is pivotal, if ongoing and emerging animal welfare issues are to be positively addressed.

For many years the veterinary profession, animal welfare organisations and compassionate individuals have worked to improve the lives of animals in many settings; including farming, working, companion, research, entertainment and animals in the wild. While a great deal has been delivered that has improved animal welfare, issues of suffering, abuse and neglect continue, with the cause in the majority of cases being the human animal's behaviour. So why is it that many of those interacting with animals, either do not follow the available advice given by veterinarians, animal behaviourists and other experts, to improve their animals life? Alternatively, in contrast, follow the advice or behaviour of people who use approaches that lead to negative outcomes for the animal and its welfare?

Taking the example dental disease in cats which can result in pain and subsequent behavioural changes (Cats Protection, 2013), it is estimated that 85% of

* Corresponding author: jo.white@progressiveideas.co.uk

cats aged three and over suffer from this preventable disease (International Cat Care, 2017). Owners are reticent to change their behaviour from buying readily available diets that they have used for years to more suitable diets and care that will promote healthy gums and teeth (Hale, 2006; Mata, 2015). It is clear that there are some factors at work creating barriers to change. For example, habitual behaviour and mental shortcuts in relation to people buying the same food, as little effort is involved. Also, the impact of social influence and norms in following what their peers, role models and the masses do. Finally, attitudes and beliefs about what is best for their cat, commonly linked to the emotions of wanting to feed the cat what the owner perceives their cat prefers. There is also change needed in the wider context regarding the approach taken by pet food manufacturers to one where they sell products aimed at promoting feline health, as opposed to focusing on what is profitable. It is clear that there is a need to be aware of the factors associated with human behaviour change (HBC) in developing future interventions, including how to successfully motivate pro-welfare action through effective communication and education about why change is required, together with the relevant support and infrastructure to facilitate the change.

It is the understanding of the barriers and the potential opportunities for change that has led to a growing interest in HBC to deliver a sustainable impact upon animal welfare. It is this understanding that has resulted in the examination of how social science, social marketing, economics, design and innovation, can be utilised to move from harmful to positive behaviours that improve animal welfare and human–animal interactions. For some time the sectors of human health, environment and economics have recognised that to deliver lasting change it is not enough to raise awareness or change attitudes, but actual sustained HBC must be delivered (Holt, 2015). It is this recognition of the need to engage HBC that has been applied to work aiming to reduce smoking, obesity, increase recycling and encourage people to save for their old age, among other examples. However, delivering measurable impact is only possible through properly understanding human behaviour in general, but also the target behaviour for change, together with the associated antecedents, drivers, barriers and opportunities.

The growing amount of evidence around the need to improve dairy cattle herd welfare (Broom, 2013; Weary and von Keyerlingk, 2017) presents a good example, as there is a recognition by those developing interventions of the need to understand why there is a problem and what its causes are. Research in the UK utilising the DairyCo Healthy Feet Programme (Reaseheath College, 2013) has shown that while all farmers in the study recognised the importance of lameness as an issue for the British Dairy Industry, tackling lameness was seen as a choice. The research indicated that the most successful farmers who engaged in the intervention were more confident in implementing change and more in control of the situation; pointing out that self-esteem and self-efficacy may have a role to play. The intervention raised awareness, knowledge and understanding of the issues and how to practically resolve them. While all farmers had similar managerial, technical and financial barriers to change; the attitude of the more successful farmers was to view the barriers to a lesser extent when compared with the least successful farmers. This work illustrates the importance of understanding the different factors involved in behaviour and attitudes, as the authors concluded that the 'ability to reduce lameness is at least in part influenced by the prevailing beliefs of those managing cows on farm' (Reaseheath College, 2013). This work also highlights the need to further understand and work towards changing the behaviour of those least successful farmers.

This talk will examine how theories and models (e.g. perception, social influence, learning, motivation, Theory of Planned Behaviour (Ajzen, 1985) and Theory of Change) can be practically applied. The purpose of implementing these models and theories is to examine the layers of an animal welfare issue regarding understanding the human behaviours that have either caused it or impacted upon it; whether these are by individual people, groups, communities, populations, or indeed, ourselves. Once the behaviours and associated factors are identified a knowledge base is developed, it is then possible to use evidence-based theories and models of HBC (e.g. Stages of Change/Transtheoretical model (Prochaska *et al.*, 1992), Theory of Interpersonal Behaviour (Triandis, 1977), Behaviour Change Wheel and COM-B Model (Michie *et al.*, 2014)) to identify behaviour change approaches, interventions, techniques and in some cases policies. The findings from the process of applying HBC models to the analysis of a real life situation, can then be incorporated into project design, and ultimately, day-to-day practical application with those directly or indirectly involved with the care or use of animals.

There is a growing number of examples where theories such as the COM-B model (Michie *et al.*, 2014) can be utilised in animal welfare (White, 2016); helping with the initial situation analysis to establish the driving factors, barriers and opportunities for change, to designing and implementing targeted intervention techniques. For example, projects looking to impact upon working animals in the developing world (The Brooke, 2017), where behavioural pain indicators are commonly observed together with physiological issues such as injuries, are applying HBC theory. Models and principles of HBC are being utilised to examine why behaviours such as working an exhausted animal, aggression in the form of beating, and the use of ill-fitting equipment that causes injuries, occur. The COM-B model aids in providing a process to gain a greater insight into the sources of the behaviour in terms of capacity, opportunity and motivation that will aid in developing successful targets for interventions, and the policy areas that are required to deliver the intervention functions. In the case of working animals, it may be that a person does not have the knowledge and skill to recognise when an animal is exhausted, as they have grown up in a culture and environment where this physiological state is the norm for both working animal and human working animal.

When looking to develop successful HBC interventions, it is important to utilise the knowledge, expertise and evidence that exists regarding intervention techniques. For example, in the case of pain-related behaviours in dogs suffering from arthritis that start to exhibit aggression (Camps *et al.*, 2012), this is not always understood or well handled by owners, who may respond by punishing the animal. However, the owner may not make the connection that the dog is in pain and therefore becoming aggressive in certain situations. Another example is the case of an overweight Labrador, while the owner may hold the attitude that being overweight is unhealthy for dogs and not good for animals suffering from conditions such as arthritis, the owner may be socially influenced by seeing other Labradors that are also overweight and move in a certain way. They then form the assumption that this is the norm for the breed, resulting in a continuation of the behaviour of overfeeding (Bland *et al.*, 2009), punishing the dog for aggression and not taking action to obtain pain-relieving drugs. With the correct HBC intervention, the owner will begin to realise and understand that the dog needs to lose weight, that the pain can be alleviated using drugs, and that the aggression is a symptom of the problem that must be tackled. However, if this is not handled correctly and an inappropriate intervention used, the owner may move into a state of cognitive dissonance and denial, making it harder to address the animal welfare issue as new barriers develop.

As professionals, veterinarians and non-veterinarians working in a pivotal role in animal behaviour and welfare, we are well placed to utilise HBC. Using HBC, we can enhance the impact and achieve greater results in day-to-day practice, as well as in the bigger picture as animal advocates (e.g. policy development and implementation, educators, role models). Examples include understanding why a client may not be compliant with advice and instructions given to improve their animal's health and welfare, and how to remedy this through correctly framing communications regarding the person's values and beliefs. How to choose the most appropriate educational techniques, including assessing when practical hands-on training is needed to develop skills, self-efficacy and confidence in a client, together with increased knowledge and awareness. It is also important to note the role of motivation in terms of ensuring that the desired behaviour is performed, together with working to develop positive routines and habitual behaviour that will deliver sustained change. Also, when to turn to other intervention approaches such as modelling, coercion or environmental restructuring for example.

The talk will provide an overview of the value of HBC to animal welfare, with insights into the theories and models, together with examples of how some of these can be applied in practice. It is the view of the organisation Human Behaviour Change for Animals that training of veterinarians and other animal experts in HBC is essential to aid in delivering positive, sustainable animal welfare, which can ultimately result in a good life for all animals.

References

Ajzen, I. (1985) From intentions to actions: a theory of planned behavior. In: Kuhl, J. and Beckmann, J. (eds) *Action Control: From Cognition to Behavior*. Springer, Berlin, pp. 11–39.

Bland, I.M., Guthrie-Jones, A., Taylor, R.D. and Hill, J. (2009) Dog obesity: owner attitudes and behaviour. *Preventive Veterinary Medicine* 1 92(4), 333–340. Available at: https://doi.org/10.1016/j (accessed 2009.08.016).

Broom, D.M. (2013) Cow welfare and herd size: towards a sustainable dairy industry. *Cattle Practice* 21(0), 169–173.

Camps, T., Amat, M., Mariotti, V.M., Le Brech, S. and Manteca, X. (2012) Pain-related aggression in dogs: 12 clinical cases. *Journal of Veterinary Behavior* 7(2), 99–102. Available at: http://dx.doi.org/10.1016/j.jveb.2011.08.002.

Cats Protection (2013) Teeth and Oral Health. *Cats Protection Veterinary Guide* 14.

Hale, F.A. (2006) Home Care for the prevention of periodontal disease in dogs and cats. Hale Veterinary Clinic. Available at: http://www.hillsvet.ca/HillsVetUS/v1/portal/en/ca/content/research/oral-health-problems/conf-pro-home-health-care.pdf (accessed 12 June 2017).

Holt, N. (2015) *Psychology: The Science of Mind and Behaviour*, 3rd edn. McGraw-Hill Education, Maidenhead, UK.

International Cat Care (2017) International Cat care. Available at: https://icatcare.org/advice/cat-health/dental-disease-cats (accessed 12 June 2017).

Mata, F. (2015) The choice of diet affects the oral health of the domestic cat. *Animals: An Open Access Journal from MDPI* 5(1), 101–109. Available at: http://doi.org/10.3390/ani5010101.

Michie, S., Atkins, L. and West, R. (2014) *The Behaviour Change Wheel: A Guide to Designing Interventions*. Silverback Publishing, Bream, UK.

Prochaska, J.O., DiClemente, C.C. and Norcross, J.C. (1992) In search of how people change: applications to addictive behaviors. *American Psychologist* 47, 1102–1114.

Reaseheath College (2013) Cattle Mobility: Changing behaviour to improve health and welfare and dairy farm businesses. Available at: http://www.reaseheath.ac.uk/wp-content/uploads/2014/02/Cattle-Mobility-Final-report-December-2013.pdf (accessed 12 June 2017).

The Brooke (2017) Thebrookeorg. Available at: https://www.thebrooke.org/about-brooke/our-strategy/how-we-do-it-our-theory-change (accessed 12 June 2017).

Triandis, H. (1977) Interpersonal Behavior, Brooks/Cole Pub. Co.

Weary, D.M. and Von Keyerlingk, M.A.G. (2017) Public concerns about dairy-cow welfare: how should the industry respond? *Animal Production Science* 57, 1201–1209.

White, J. (2016) Key Principles of Human Behaviour Change. Proceedings of the 1st International Conference on Human Behaviour Change for Animal Welfare, Dorking, UK.

Impact of Exploratory Material and Stocking Density on Tail and Ear Biting in Suckling and Weaning Piglets

CHRISTINE LEEB*, KERSTIN APER AND CHRISTOPH WINCKLER

University of Natural Resources and Life Sciences (BOKU), Vienna, Austria

Conflict of interest: The authors declare no conflict of interest.

Keywords: piglets, exploration, density, tail-biting, ear-biting

Introduction

Increasingly, farmers and retailers are working towards solutions to keep pigs with intact tails. Therefore, the aim of this study was to investigate the impact of additional material (hay in a rack additional to sisal rope) on tail and ear biting in suckling and weaned piglets. In addition, the effect of reduced stocking density was examined for weaners.

Materials and Methods

Three batches (each six litters) of undocked piglets were included. Suckling piglets received either a rope (C) or additionally hay in a rack (H). Weaners were kept with the same type of exploratory material allocated to groups with normal (N) (0.35 m²/pig) or reduced (R) stocking density (0.50 m²/pig), resulting per batch in one group with each factor combination. Continuous behaviour sampling using direct observations twice weekly was carried out including two segments each of 10 minutes (piglets) and 15 minutes (weaners) per pen. Data were analysed using linear mixed models.

* Corresponding author: Christine.leeb@boku.ac.at

Results

Suckling piglets in groups H showed fewer manipulations of tails, ears and body, and more exploration of material. In weaner groups H, ear manipulation was significantly lower; other behaviours were neither affected by material nor stocking density. However, eight weaners (all C) had to be removed as biters within the first two batches.

Discussion

Sufficient quality and quantity of exploratory material are essential during the suckling period as behavioural differences were observed at this early age. Furthermore, observation of animals together with management of material to ensure constant availability are the basis to keep pigs with intact tails.

Ontogeny of Selected Behaviours in Piglets of Slovak Large White Improved Swine Breed

LENKA LEŠKOVÁ*, ANDREA JURKOVÁ AND JANA KOTTFEROVÁ

University of Veterinary Medicine and Pharmacy in Košice, Košice, Slovakia

Conflict of interest: The authors declare no conflict of interest.

Keywords: behaviour, ontogeny, piglets

Introduction

Industrial farming helps meet the growing needs of the population. However, it also leads to numerous concerns regarding animal welfare. Various behaviour problems may be observed due to restricted movement, feeding and lack of stimuli. The aim of this study was to evaluate changes in the behaviour of piglets during their development from birth to weaning under extensive conditions.

Materials and Methods

A litter of five females and four males of Slovak Large White Improved swine breed was evaluated. The sow and piglets were kept in extensive conditions (deep straw bedding and no restriction of movement). The piglets had *ad libitum* contact with the sow. The behaviour of the piglets was recorded by a video camera and evaluated using the Noldus Observer XT Analysis software. Observations were carried out on a weekly basis in the first 5 weeks post-partum. Events such as feeding, sleeping, rooting, play and agonistic interactions were recorded. The percentage of time spent on each behaviour was noted for each week of age.

* Corresponding author: lenka.leskova@uvlf.sk

Results

In the first week post-partum, piglets mostly suckled (45.7%) and slept (38.5%). During the second week, rooting started appearing at a minimal level (0.5%), but by the fifth week of age, it increased significantly (29.5%). Play and agonistic interactions were the following behaviours to appear, from the third week of age. An offensive threat appeared in the third week (0.3%), while defensive behaviour was first observed during the fifth week of age (0.1%).

Conclusion

The result of this pilot study showed that sows and piglets in extensive conditions thrive. Agonistic behaviours that can lead to future production losses were kept to a minimum. Moreover, rooting and play were prevalent and increased with age. Further studies on larger scales (e.g. industrial pig farms) are needed to evaluate these findings.

Access to Chewable Materials Increases Piglet Activity During Lactation

Kirsi-Marja Swan*, Helena Telkänranta, Camilla Munsterhjelm, Olli Peltoniemi and Anna Valros

University of Helsinki, Department of Production Animal Medicine, Helsinki, Finland

Conflict of interest: The authors declare no conflict of interest.

Keywords: piglet, behaviour, enrichment, sow, lactation

Introduction

Studies have shown the benefits of enrichment after weaning, yet, there is little data on the influence during lactation. The aim was to investigate the effects of enrichment on activity levels.

Materials and Methods

Litters of 29 sows had access to ten sisal ropes, plastic ball, newspaper and wood shavings (the rope-paper group, RP) (n=29). The control group (C), had access to wood shavings and plastic ball (n=29). Behaviours were recorded for 4-hour periods. We analysed the behaviour of the piglets, successful nursing events and behaviour of the sows. Mann–Whitney U-test was done for two age groups: 7–13 days old (GROUP 1) (n(RP)=22, n(C)=22); and 14 days or older (GROUP 2) (n(RP)=24, n(C)=24).

* Corresponding author: kirsi.swan@helsinki.fi

Results

In GROUP 1 the RP piglets manipulated the udder longer (p=0.023; median 83 minutes, range 30–119) and more often (p<0.001; 91 events, 39–145) than group (C) (49 min, 20–109; 62 events, 34–154). In GROUP 2 the RP piglets made body contacts towards the sow more often (p=0.047; median 124 events, 41–178), than group (C) (52 events, 12–116). The sows were standing (p=0.01) and performing oral–nasal manipulation (p<0.01) more often in group (C) (median four events, 0–11 and 15 events, 2–31) than in the RP group (two events, 0–9 and five events, 0–28). The duration of standing (p=0.02) and oral–nasal manipulation (p=0.02) was longer in group (C) (17 minutes, 0–68; 11 minutes, 1–50) than in the RP group (6 minutes, 0–47; 3 minutes, 0–33).

Conclusion

Enrichment increases the activity of the piglets, which may affect the sow–piglet interaction, sow behaviour and lactation performance.

Hypothermia Triggers Depression-like Behaviour in Mice Forced Swimming Test

ÇIGDEM ALTINSAAT[1]*, HASAN ÇALIŞKAN[2], NESRIN SULU[1] AND ETKIN ŞAFAK[1]

[1]Department of Physiology, Faculty of Veterinary Medicine, Ankara University, Turkey; [2]Department of Physiology, Faculty of Medicine, Ankara University, Turkey

Conflict of interest: The authors declare no conflict of interest.

Keywords: depression, forced swim test, hypothermia, mice

Introduction

Depression is a prevalent public health concern. Different environmental factors such as hypothermia, hypoxia and mid-season nutritional deficiency can cause depression-like behaviour in animals. The aim of this study was to investigate the effect of hypothermia on depression-like behaviour.

Materials and Methods

Forty-two male mice, 10–12 weeks old were used in this study. All subjects were fed *ad libitum* during the experiment. Subjects were randomly divided into seven groups (n=6) based on water temperature (8, 11, 14, 17, 20, 23 and 26°C). Classical Forced-Swimming Test (FST) protocol was selected and each lasted 6 minutes; however, only the last four were evaluated (Porsolt *et al.*, 1977). Climbing, swimming, floating and latency time were analysed using one-way ANOVA and then Tukey's post-hoc test.

* Corresponding author: caltinsaat@yahoo.com

Results

Floating times increased significantly in 8, 11 and 14°C when compared to 17, 20, 23 and 26°C (p<0.01). Swimming times reduced in 8, 11, 14 and 17°C when compared to 23 and 26°C. Mice increased swimming in 20°C compared to 8°C (p<0.05). Latency times decreased in 8, 11, 14 and 17°C when compared to 20, 23 and 26°C (p<0.05). Climbing times in 17°C and 20°C increased significantly. Climbing reduced remarkably in 8, 11, 14, 23 and 26°C when compared to 17 and 20°C (p<0.05).

Discussion

Our results suggest that hypothermia increased floating time, indicating depression-like behaviour. Escape behaviours including swimming and climbing were reduced as response to hypothermia. Our data suggest that adjusting water tempature during FST protocol is important to avoid hypothermia which causes depression-like behavior.

Acknowledgement

This research received ethical approval (Protocol No: 2015-7-115) from Ethics Committee of Ankara University.

References

Porsolt, R.D., Bertin, A. and Jalfre, M. (1977) Behavioral despair in mice: a primary screening test for antidepressants. *Archives internationales de pharmacodynamie et de therapie* 229, 327–336.

Effects of Maternal Depression and Antidepressant Therapy on the Neurobehavioural Development of Rat Offspring

Michal Dubovicky*, Eszter Bögi, Kristína Belovičová, Romana Koprdová and Mojmír Mach

Institute of Experimental Pharmacology and Toxicology, Slovak Academy of Sciences, Bratislava, Slovakia

Conflict of interest: The authors declare no conflict of interest.

Keywords: maternal depression, treatment, rat, pregnancy, behaviour

Introduction

Depression during pregnancy and in the post-partum period is a growing health concern in modern society. The use of psychopharmacological agents during gestation and lactation raises several questions, which are mainly related to safety during these periods. The main concern is that antidepressants cross the blood–brain, placental barrier and are excreted in the milk and may directly affect fetal and neonatal development of the offspring. Antidepressant treatment during pregnancy could interfere with sensitive developmental processes in the brain and have neurobehavioural consequences in later life. The aim of this study was to investigate the effects of venlafaxine on selected neurobehavioural variables in mothers and their offspring.

Material and Method

Stressed and non-stressed Wistar rat dams were treated with either venlafaxine (10 mg/kg/day) or placebo. Using a model of maternal adversity the dams and

---

* Corresponding author: michal.dubovicky@savba.sk

pups were evaluated. Hormonal and metabolic levels and behaviour of the rats were measured.

Results

The results show that administration of venlafaxine during gestation and lactation did not affect different biochemical, reproductive and endocrine variables of the rat offspring nor affect play behaviour of the rat pups in their home environment. Adolescent offspring showed altered behaviour in a new and stressful environment.

Discussion

These results suggest that exposure to venlafaxine during gestation and lactation could interfere with the functional development of the brain and can cause behavioural changes on the level of adaptation of animals. These changes may be visible in the neonatal period. However, they may appear only during adolescence or adulthood in a new stressful situation.

From Neuronal Activity to Behaviour: Understanding Neuronal Correlates of Sensory Discrimination

Tomas Hromadka[1,2]*

[1]*Institute of Neuroimmunology, Slovak Academy of Sciences, Bratislava, Slovakia;* [2]*Axon Neuroscience R&D Services, Bratislava, Slovakia*

Conflict of interest: The author declares no conflict of interest.

Keywords: Alzheimer's, neurodegeneration, cognitive deficits, sensory deficits

Alzheimer's disease is currently the most common neurodegenerative disorder, characterised by distinct cognitive and sensory deficits. The underlying pathogenic mechanism, however, remains elusive. How the distinct molecular and morphological changes affect information processing in neuronal circuits and translate further into cognitive dysfunction is unclear. Resolving the changes in neuronal activity is crucial for understanding the impact of neurodegeneration on brain function and even for possible identification of effective treatments and diagnostic techniques.

The progression of neurodegeneration is associated with complex changes in overall neuronal activity, such as the imbalance between excitatory and inhibitory transmission. Dysfunction of GABAergic transmission has emerged recently as one of the key players in the pathogenesis of network dysfunction in Alzheimer's disease. It is, however, still unclear which types of neurons and neuronal assemblies are most affected by the constant process of neurodegeneration. Pathological activity and function of distinct neuronal classes reverberates across neuronal networks and likely causes an imbalance in the overall activity of cortical networks leading to cognitive deficits characteristic of Alzheimer's disease.

We use a combination of electrophysiological, optical, molecular and optogenetic tools *in vivo* to study the impact of neurodegeneration at the level of

* tomas.hromadka@savba.sk

individual well-defined classes of cortical interneurons and their neuronal circuits in sensory and association cortical areas. We also strive to identify the changing role of cortical interneurons and their microcircuits in mediating cognitive processes during neurodegeneration. Such functional organisation of interneurons would directly relate the detailed structure of cortical circuits to perception and behaviour in brain disorders. Detailed knowledge of deviations from normal information processing in diseased brains will form a platform for seeking novel therapeutic and diagnostic strategies even in the initial presymptomatic stages of Alzheimer's disease.

Environmental Control of Crib-biting in a Horse

MICHAELA HEMPEN[1]* AND JESÚS ROSALES-RUIZ[2]

[1]University of Edinburgh, Royal (Dick) School of Veterinary Studies, Roslin, UK; [2]University of North Texas, Department of Behavior Analysis, Denton, Texas, USA

Conflict of interest: The authors declare no conflict of interest.

Keywords: crib-biting, equine, stereotypy, behaviour, applied behaviour analysis, stimulus control analysis of behaviour

Introduction

The most common stereotypic behaviours in horses are crib-biting/wind-sucking, weaving and box-walking. The reported prevalence for crib-biting in horses is 2.4–8.3%. Crib-biting is most frequently reported to occur during and following the consumption of meals and has been linked to clinical conditions. Current approaches to prevent or reduce crib-biting are often not successful. This study assessed whether previously researched strategies from the field of applied behaviour analysis could be used to assess the stimulus control for crib-biting in horses and influence when crib-biting occurred.

Materials and Methods

Two stabled horses were observed for 24 hours to establish a pattern in their crib-biting behaviour. An analysis was conducted for one horse to determine the stimulus conditions that were likely to evoke the crib-biting behaviour. Based on the outcome of this analysis, a treatment protocol was applied to see if crib-biting could be directly evoked or interfered.

* Corresponding author: michaelahempen@gmail.com

Results

The results from the first part of the study showed that crib-biting patterns could be altered by changing the time at which the horse was fed. The results from the second part of the study showed that a piece of carrot could reliably evoke crib-biting and that contingent tactile reinforcers could reduce crib-biting to zero.

Conclusion

This study provides evidence that crib-biting in horses can be disrupted by changing the stimulus conditions that predict reinforcement. This opens the possibility of treatment options based on operant conditioning techniques and the use of stimulus control to reduce the frequency of the behaviour.

Do Different Types of Food Provoke Different Levels of Enjoyment in Dogs?

Franck Péron[1]*, Julien Roguès[1], Caroline Gilbert[2] and Emira Mehinagic[1]

[1]Diana Pet Food, Elven, France; [2]Université Paris-Est, Ecole Nationale Vétérinaire d'Alfort, Maisons-Alfort, France

Conflict of interest: Franck Peron, Julien Roguès and Emira Mehinagic are paid employees of Diana Pet Food.

Keywords: emotion, dogs, arousal, heart rate variability, thermography, behaviour, food

Introduction

The aim of our study was to investigate several behavioural and physiological parameters to assess positive emotions while dogs were experiencing different food stimuli.

Material and Methods

Thirty-nine young adult dogs, males and females, from different breeds were involved in a two-step study. Each dog received different sessions all organised the same: (P1) 5 minutes baseline; (P2) distribution of one piece of food every 20 seconds for 5 minutes; and (P3) 5 minutes post-ingestive period. In phase one, effects of presenting cooked pasta and treats were compared looking at different behavioural criteria (e.g. postures, transitions and tail wagging), eye temperature (Fluke TiR1 thermal imager) and heart rate variability (HRV) parameters (Polar ©). In phase two, HRV was measured while dogs received two types of kibbles differing only in their percentage of palatability enhancer coating. Mixed-effect model for each variable was performed including the dog as random effect,

* Corresponding author: fperon@diana-petfood.com

the baseline as a fixed quantitative effect, and the product, the period and the order as fixed qualitative effects. All the effects were tested at 5%.

Results

Dogs display different pattern of behaviours while receiving different products. The eye temperature did not alter significantly between the food stimuli while the HRV parameters (SDNN, RMSSD and pNN50) changed. Parameters showed increased arousal for highly palatable foods.

Conclusion

Not only food distribution provokes positive arousal but also variations of the stimuli quality are observed at subtler physiological level suggesting that dogs enjoy differently different food products.

Bibliography

Kostarczyk, E. and Fonberg, E. (1982) Characteristics of the heart rate in relation to the palatability of food in dogs. *Appetite* 3(4), 321–328.
Travain, T., Colombo, E.S., Grandi, L.C., Heinzl, E., Pelosi, A., Previde, E.P. and Valsecchi, P. (2016) How good is this food? a study on dogs' emotional responses to a potentially pleasant event using infrared thermography. *Physiology & Behavior* 159, 80–87.
Zupan, M., Buskas, J., Altimiras, J. and Keeling, L.J. (2016) Assessing positive emotional states in dogs using heart rate and heart rate variability. *Physiology & Behavior* 155, 102–111.

Survey on Shock Collar Use in France: Providing Practical Results for Regulatory Guidelines Development

SYLVIA MASSON[1]*, ISABELLE NIGRON[2]
AND EMMANUEL GAULTIER[3]

[1]Clinique de la Tivollière, Voreppe, France; [2]Clinique vétérinaire Nigron André, Roanne, France; [3]FERCEA, Cabrieres d'Avignon, France

Conflict of interest: The authors declare no conflict of interest.

Keywords: e-collar, dog training, punishment, shock, dog welfare

Introduction

Electronic collar training is controversial and current legislation regarding e-collars varies from no regulation to a complete ban across Europe. The main goal of this study is to characterise the practical use of such tools in France, where no regulation exists, and to compare it to theoretical experimental data, to provide accurate information for a potential future regulation. Additionally, a study on the differences between types of collars was conducted to investigate potential nuances.

Material and Methods

A sample of dog owners (n=1251) was recruited using an online questionnaire. Data were collected during two months using Google Forms. Factors associated with e-collar use were determined using the chi-squared test and Fisher's exact test.

* Corresponding author: s.masson@hotmail.com

Results

Approximately 26% (n=330) of the owners enrolled in this survey used electronic devices; 11.9% (n=149) of owners reported the use of bark-activated collars, 4.5% (n=56) for electronic boundary fences, and 14.2% (n=178) for remote-controlled collars. Three factors were found to be significantly correlated to e-collar use: high weight dog ($\chi^2 = 18.5$, df = 7, p < 0.001), non-neutered status ($\chi^2 = 6.05$, df = 1, p < 0.02) and the nature of discipline practised with the dog ($\chi^2 = 81.9$, df = 6, p < 0.001). The data collected showed minimal guidance before use (<29%) on young dogs (<2 years) and without trying other methods beforehand. Forty-six per cent of owners using an e-collar indicated that they use rewards as well. Seven per cent of the dogs on which an e-collar has been used presented physical burns (n=23). The efficacy reported (<50%) was lower than the efficacy reported by professional trainers (90%) (Cooper *et al.* 2014). Bark-activated collars were used mainly due to neighbours' complaints; electronic boundary fence collars were used mostly where no physical barrier exist; and remote-activated collars were mostly used when dogs did not respond to a recall cue. Bark-activated collars appeared to be the least efficient (25.5% success), used mostly on dogs that showed abnormal behaviours beforehand ($\chi^2 = 38.5$, df = 1, p <0.001), and caused most injuries (10.7%; p=0.0033). Remote-activated collars were used mostly for owners' comfort and like, bark-activated ones, its users describe more abnormal behaviours beforehand ($\chi^2 = 20.7$, df = 1, p <0.001) when compared to non-users.

After using collars, owners describe their dogs' behaviour as less excited, more calm or more sad, which coincides with the scientific data describing dogs with a general behavioural inhibition after an e-collar has been used (Schilder and Van Der Borg, 2004).

Conclusion

The results of this survey indicate a frequent use of e-collars in France. Moreover, there is a lack of knowledge of proper application of e-collars that may lead to reduced welfare of dogs. Nearly half of the owners in this study indicated that they use rewards alongside using e-collars. This may indicate that these owners do not perceive e-collars as a punishment. In addition, these data show the differences between e-collar types that should be further investigated to allow for more precise regulation. Finally, due to potential danger in using these e-collars, there is an urgent need to regulate the use of this tool in dog training.

References

Cooper, J.J., Cracknell, N., Hardiman, J., Wright, H. and Mills, D. (2014) The welfare consequences and efficacy of training pet dogs with remote electronic training collars in comparison to reward based training. *PLoS One* 9, DOI: 10.1371/journal.pone.0102722.

Schilder, M.B.H. and Van Der Borg, J.A.M. (2004) Training dogs with help of the shock collar: short and long term behavioural effects. *Applied Animal Behaviour Science* 85, 319–334, DOI:10.1016/j.applanim.2003.10.004.

'Long-stay Dogs' in Shelters: Studying Factors Related to Adoptability of Difficult to Adopt Dogs in Catalonia

Marta Calcerrada[1]*, Lorena Torre[1], Paula Calvo[1], Jonathan Bowen[1,2] and Jaume Fatjó[1]

[1]Chair Affinity Foundation Animals and Health, Universitat Autònoma de Barcelona, Spain; [2]Royal Veterinary College, University of London, UK

Conflict of interest: The authors declare no conflict of interest.

Keywords: long-stay, dogs, adoption, shelter, behaviour problems

Introduction

Animal shelters usually house 'long-stay dogs', whose characteristics make their adoption difficult. Their main profiles include: dangerous (under Spanish legislation), senior, chronic disease or with behavioural problems. Our aim was to explore the motivations of potential adopters when choosing these dogs.

Material and Methods

An online survey investigated general motivations for dog adoption and responsible ownership. Participants were recruited via email campaigns from three shelters in Catalonia. Kruskal–Wallis (KW), and Mann–Whitney U tests were performed to compare scoring for dog profiles and analyse factors related to motivations to adopting each.

---

* Corresponding author: catedra.animales.salud@gmail.com

Results

We obtained 1336 complete answers. Ninety-three per cent of respondents indicated that they would adopt a dog in the future. Regarding the motivation to adopt, senior dogs scored the highest and dogs with behaviour problems scored the lowest (KW test, with Dunn's multiple comparisons). Women scored significantly higher than men on motivation to adopt any long-stay dogs. Young people (aged 18–25) were significantly more motivated to adopt dangerous dogs and less motivated to adopt senior dogs (KW test, with Dunn's multiple comparisons). Most frequent objections to adopting a long-stay dog were 'unable to afford it' and 'lack of time'. Seventy-two per cent of respondents agreed they would adopt a long-stay dog if shelters offered veterinary support or behavioural treatment.

Conclusion

The results suggest that characteristics of potential adopters, such as gender and age, should be considered for adoption campaigns of long-stay dogs. Behaviour problems appear to deter adopters, and behavioural support from shelters seems to be required.

Acknowledgement

The present study was developed with the support of Affinity Foundation and the Barcelona City Council.

Compulsive Self-licking and Self-biting in Dogs with Paraesthesia: Two Cases

Carlo Siracusa*, Lena Provoost, M. Leanne Lilly, Lili Duda, Leontine Benedicenti and Ludovica Chiavaccini

Department of Clinical Studies, School of Veterinary Medicine, University of Pennsylvania, Philadelphia, Pennsylvania, USA

Conflict of interest: The authors declare no conflict of interest.

Keywords: compulsive behavior, dog, paraesthesia

Case One

History and presenting signs

A 10.5-year-old castrated male Coonhound/Great Dane cross presented with a 1-year history of anxious behaviour and repetitive licking of the left carpus, which started following a 3-week absence of the owners. They noted a small button-like tumour on the left carpus, which had grown to 8×8 cm size at presentation.

Diagnosis

The repetitive licking was thought to be initiated by separation anxiety and perpetuated by discomfort/pain.

Treatment

Treatment included behaviour modification, amitriptyline, gabapentin and amantadine. Histopathology revealed a fibroadnexal hamartoma. Palliative

---

* Corresponding author: siracusa@vet.upenn.edu

radiotherapy resulted in tumour shrinkage but no change in licking. Ultrasound imaging revealed entrapment of the ulnar nerves as the suspected cause of paraesthesia (Hankins, 2012). Locoregional anaesthesia of the radial and dorsal ulnar nerves with bupivacaine, dexmedetomidine and triamcinolone stopped licking for 2 weeks and decreased licking frequency for another 2 weeks. The owners elected to repeat locoregional anaesthesia as needed.

Case Two

History and presenting signs

A 3-year-old castrated male Golden Retriever/Poodle cross presented for repetitive licking/biting of the tail base and right thigh, walking and whimpering while defaecating, pressing his hindquarters against the owner, and anxious/fearful behaviour. Licking started at 14 months of age, and intensified into biting, causing open wounds. At presentation, MRI revealed a synovial cyst at the right L5–L6 articular process and a Cd2–Cd3 vertebral sclerosis and narrowing of the canal.

Diagnosis

Dynamic compression of the correspondent nerves was the suspected cause of paraesthesia at the tail and thigh.

Treatment

Behaviour modification, clomipramine, gabapentin and amantadine decreased the frequency but not intensity of the compulsive behaviour. Epidural infiltration of bupivacaine/triamcinolone (Janssens et al., 2009) temporarily improved the repetitive behavior.

References

Hankins, C.L. (2012) Carpal tunnel syndrome caused by a fibrolipomatous hamartoma of the median nerve treated by endoscopic release of the carpal tunnel. *Journal of Plastic Surgery and Hand Surgery* 46, 124–127.

Janssens, L., Beosier, Y. and Daems, R. (2009) Lumbosacral degenerative stenosis in the dog. The results of epidural infiltration with methylprednisolone acetate: a retrospective study. *Veterinary and Comparative Orthopaedics and Traumatology* 6, 486–491.

Use of Huperzine A in Three Canine Cases in Australia

KERSTI SEKSEL* AND GRACE THURTELL

Sydney Animal Behaviour Service, Seaforth, New South Wales, Australia

Conflict of interest: The authors declare no conflict of interest.

Keywords: dog, huperzine, anxiety

Case One

History and presenting signs

A 5-year-old neutered female American Staffordshire Bull Terrier cross weighing 22 kg was presented for excessive vocalisation and attention seeking as well as aggression directed towards unfamiliar people.

Diagnosis

An anxiety disorder was diagnosed based on clinical signs and history. Differential diagnoses were ruled out by the referring behaviour veterinarian based on physical examination and evaluation of the dog's physical and social environment.

Treatment

The treatment included environmental management, behaviour modification and medication. The dog was prescribed fluoxetine and amitriptyline initially. When the dog's anxiety was not reduced, huperzine was added. Huperzine, a Chinese club moss alkaloid, is an acetylcholinesterase inhibitor and a neuroprotective agent with an N-methyl-D-aspartate agonist that may have potential benefits in cases of cognitive impairment and depressive disorders (Tun and Herzon, 2012).

* Corresponding author: sabs@sabs.com.au

The introduction of huperzine led to a noticeable increase in the dog's general reactivity and barking. Once huperzine was stopped her behaviour seemed to improve quickly and barking and reactivity reduced, with a clear change back to how she was before starting the huperzine.

Case Two

History and presenting signs

A 6-year-old neutered male Shiba-Inu weighing 15 kg was presented for separation distress, aggression directed towards unfamiliar dogs and the tea drawer in the kitchen.

Diagnosis

An anxiety disorder was diagnosed based on clinical signs and history. Other differentials were ruled out based on history and evaluation of the dog's physical and social environment.

Treatment

The treatment included environmental management, behaviour modification and medication. The dog was prescribed clomipramine, trazodone and clonidine. Huperzine was added later. The owners reported they felt that the dog's arousal and aggression levels became more generalised and increased in intensity. Huperzine was discontinued and the dog's behaviour improved.

Case Three

History and presenting signs

A 3-year-old neutered male Labrador Retriever weighing 25.9 kg was presented for anxiety and a possible compulsive disorder.

Diagnosis

An anxiety disorder was diagnosed based on history and clinical signs. The differential diagnoses such as lack of stimulation, pain, and medical conditions such as skin and gastrointestinal disorders were ruled out by the referring behaviour veterinarian through full physical examination, and evaluation of the dog's physical and social environment.

Treatment

Treatment included environmental management, behaviour modification and medication. The dog was prescribed clomipramine and Adaptil. Huperzine was added later. Huperzine led to an increase in the dog's agitation and aggression. The dog could not settle; he was barking and biting his owner, and he became uncontrollable. Huperzine was discontinued and eventually the dog's behaviour returned to a manageable level.

References

Tun, M. and Herzon, S. (2012) The pharmacology and therapeutic potential of (-)-huperzine A. *Journal of Experimental Pharmacology* 4, 13–123.

Mirtazapine as a Potential Drug to Treat Social Fears in Dogs: Five Case Examples

JUAN ARGÜELLES[1]*, JESÚS ENRIQUEZ[1], JON BOWEN[2] AND JAUME FATJÓ[3]

[1]Ethoclinic Valencia, Behaviour Medicine Reference Service, Valencia, Spain; [2]Queen Mother Hospital for Small Animals, The Royal Veterinary College, Hatfield, UK; [3]Departament of Psychiatry, School of Medicine, Autonomous University of Barcelona, Barcelona, Spain

Conflict of interest: The authors declare no conflict of interest.

Keywords: mirtazapine, fear, dogs

Introduction

Dogs who are fearful of people often show a reduced motivation to take food in threatening scenarios and during behaviour modification sessions. Benzodiazepines interfere with learning (Crowell-Davis and Murray, 2006) and SSRIs can further decrease appetite (Fitzgerald and Bronstein, 2013), thus limiting reward-based training. Mirtazapine has anxiolytic and appetite stimulant properties (Davis and Wilde, 1996; Nobukazu *et al.*, 2009; Quimby and Lunn, 2013). The aim of this study was to evaluate the efficacy of mirtazapine as a second-choice drug to treat social phobia in dogs when the motivation to take food is limited.

Materials and Methods

Five dogs showing fearful behaviour towards people were selected. None of them accepted food from either the behaviourist or the caregiver during training. A thorough behavioural and physical examination was performed, including haematology and biochemistry. The diagnosis was established based on context

* Corresponding author: juanin73@gmail.com

and body language. Three dogs were previously treated with fluoxetine and trazodone without an improvement. Dogs were administered mirtazapine (1–1.5 mg/kg PO sid) (Giorgi and Yun, 2012). Clinical signs and performance during training sessions were evaluated after 30 days of treatment.

Results

All patients showed clinical improvement: they were more relaxed during walks, accepted more contact from unknown people and showed fewer attempts to escape. During behaviour modification sessions, the dogs accepted treats more readily when treated with mirtazapine.

Discussion

Although commonly used in human medicine, mirtazapine is not usually mentioned in the veterinary behaviour medicine literature. The results of this study suggest that mirtazapine may be useful as a second-choice treatment for fear-related behaviours in dogs, particularly when the motivation to take food is low.

References

Crowell-Davis, S. and Murray, T. (2006) *Veterinary Psychopharmacology*. Blackwell Publishing, Ames, Iowa, pp. 34–71.

Davis, R. and Wilde, M.I. (1996) Mirtazapine: a review of its pharmacology and therapeutic potential in the management of major depression. *CNS Drugs* 5(5), 389–402.

Fitzgerald, K.T. and Bronstein, A.C. (2013) Selective serotonin reuptake inhibitor exposure. *Topics in Companion Animal Medicine* 28, 13–17.

Giorgi, M. and Yun, H. (2012) Pharmacokinetics of mirtazapine and its main metabolites in Beagle dogs: a pilot study. *The Veterinary Journal* 192, 239–241.

Nobukazu, K., Fumikazu, Y., Miki, Y., Koichi, K., Taiichiro, I., Takeshi, I. and Tsukasa, K. (2009) Anxiolytic-like profile of mirtazapine in rat conditioned fear stress model. *Pharmacology, Biochemistry and Behavior* 92, 393–398.

Quimby, J.M. and Lunn, K.F. (2013) Mirtazapine as an appetite stimulant and anti-emetic in cats with chronic kidney disease: a masked placebo-controlled crossover clinical trial. *The Veterinary Journal* 197, 651–655.

Polyuria and Polydipsia Associated with Hypersensitivity–Hyperactivity Syndrome (Attention Deficit Hyperactivity Disorder) in a Dog: A Case Report of a Male French Shepherd Dog (Beauceron)

Nathalie Marlois[1]* and Claude Béata[2]

[1]Clinique vétérinaire de l'Albarine, Ambérieu en Bugey France; [2]Toulon, France

Conflict of interest: The authors declare no conflict of interest.

Keywords: behaviour disorder, dog, hypersensitivity–hyperactivity (HSHA), ADHD, polyuria/polydipsia, fluoxetine

History and Presenting Signs

A 17-month-old neutered male Beauceron weighing 39 kg was referred for behavioural consultation. The dog presented multiple abnormalities such as polyuria and polydipsia (PUPD) with house soiling, inappropriate urination, lack of satiety, attention deficit, hypersensitivity (the dog was highly reactive to any stimuli, including benign triggers), hyperactivity, impulsivity, hyposomnia and destructive behaviours. The symptoms were observed from the time of adoption at 2 months of age.

The dog started showing food-related aggression, or when confronted, at the time of sexual maturity. Additionally, the dog started showing aggressive behaviours towards a female dog in the house.

At 6 months of age, the dog exhibited epileptic seizures and was prescribed phenobarbital. A low dosage (0.64 mg/kg PO bid) was enough to control the crises during the week. He needed more on weekends (0.96 mg/kg PO bid) to manage

* Corresponding author: nmarlois@club-internet.fr

a higher level of stimulation when attending training and to interact with other dogs. The polydipsia was present before initiation of phenobarbital therapy, and the treatment had little effect on it.

The dog lived in a house environment, and had a female owner who worked at home. The owner's elderly parents lived on the top floor of the house, and her brother came once daily at mid-day. There were two other dogs (Beauceron) in the home; two neutered females aged 10.5 and 9 years. The dogs were walked every day alone or together for several hours, with or without a leash, in town and the countryside. The owner does off-road running with the patient, but at four o'clock in the morning for safety reasons.

The owner is taking him to dog training classes that use positive methods, with many treats. At the time of presentation, this situation exhausted the owner; the patient was difficult to control in many situations, urinates constantly, is destructive, does not sleep at night, shows aggression and inappropriate interactions towards the female dogs he lives with, and has become dangerous. The dog's behaviour is consistent regardless of the people around him.

Clinical Examination

The dog was very excitable from the outset (in the waiting area and the consultation room). Aside from the patient, the owner brought her brother and the other two female dogs. During the consultation, the dog was not able to settle and rest. He constantly panted, paced, sought attention and looked for something to chew. Moreover, he would often urinate in the consultation room, especially when any contact was made with him. Finally, he tried pulling the other dogs' collars. If no attention was given, he was slightly easier to manage. His exploration was disorganised.

Clinical examination was difficult; the patient resisted restraint and urinated frequently. Physiological parameters were within normal limits; however, it was difficult to record these accurately. The dog did not exhibit any signs of pain, and no neurological signs were found.

Diagnosis

Since adoption at 2 months of age, the dog has exhibited the following symptoms:

1. Lack of biting inhibition.
2. Inability to stop a behavioural sequence or an immediate start of a new sequence. For example, hyperactivity despite adequate exercise, oral exploration leading to damage, and erratic exploration.
3. Hypersensitivity: he reacts to any stimulus, even those that are continuously present in his environment.
4. Lack of satiety: he tries anything possible to reach his food bag and could eat it all, despite being fed three times a day. He is obsessed with human food and must be locked in a kennel during family meals.
5. Hyposomnia: sleep duration is less than 8 hours a day.

These five symptoms are the historical core diagnostic criteria of stage 2 HSHA syndrome (Pageat, 1998). The dog also presented the following behaviours, which often relate to HSHA, and are increasingly considered even though they are not mandatory for diagnosis:

1. Attention deficit.
2. Hyperreactivity: a reaction to any stimulus in an exaggerated manner.
3. Impulsivity.

In addition, the dog exhibits inappropriate urination that could be due to immaturity, PUPD, increased urine osmolarity and hyperreactivity. There were no signs of incontinence.

The dog has exhibited the following behaviours since puberty:

1. Aggression because of frustration or irritation, and impulsivity. Moreover, it may be associated with impaired communication and anxiety. This aggression was reinforced through the consequences of this behaviour.
2. Seizures increased by excitation but controlled with low doses of phenobarbital.

The PUPD was investigated by the referring veterinarian, once when the dog was a puppy and once more after the behavioural consultation. An extensive diagnostic investigation was carried out by the Orbio laboratory, and the results were normal. This investigation included urinalysis, RPCU, full blood count, and biochemical and endocrine assessment (T4, TSH, fructosamine, cortisol with stimulation, RCCU, bile acid) (Schoeman, 2008). Phenobarbital serum level was 9.84 mg/l, which is very low. Ultrasound scan of the abdomen was normal. Water deprivation test (both as a puppy and as an adult) showed that the dog's urine became concentrated in response to dehydration.

We hypothesised that the polydipsia is related to the HSHA syndrome and that the seizures are linked to a high level of excitability and a very low capacity for self-control.

Differential Diagnosis

The dog exhibits the criteria for stage 2 hypersensitivity–hyperactivity (HSHA) syndrome (Pageat, 1998; Mege *et al.*, 2003). Hypersensitivity–hyperactivity in dogs is similar to human attention deficit hyperactivity disorder (ADHD) (Marlois, 2001). Increasingly, studies suggest that dogs with ADHD-like symptoms may be an interesting model for human ADHD (Lit *et al.*, 2010; Puurunen *et al.*, 2016), even though the recognition of the pathology does not have a consensus. The traditional diagnosis of hyperkinesis (Campbell, 1973) in behavioural medicine often entails physiological changes like sustained tachycardia, persistent hyperpnoea, excessive salivation, increased energy metabolism and decreased urination. It is uncommon to observe these signs together (Luescher, 1993). These signs are usually not persistent, and may not even be relevant.

Response to methylphenidate supports the diagnosis. Unfortunately, methylphenidate is not accessible to veterinarians in France. The use of methylphenidate in dogs is not always successful, and some dogs respond poorly (Luescher, 1993). In France, the diagnosis is based exclusively on the clinical signs and the behavioural and medical history.

Another consideration is the dog's primary needs. The owner provided the dog with all its biological requirements (e.g., food, water, exercise and opportunities to eliminate) and behavioural needs (training, affection, play and attention). Moreover, the owner regularly allowed her dog to interact with other dogs and people.

Congenital cerebral dysfunction may be suspected. Aside from the seizures, the dog did not show any neurological signs. Further, his ability to learn was good. Although magnetic resonance imaging (MRI) would be beneficial, it was not performed.

Management

Due to the dog's psychopathological condition and the severity of the symptoms, it was decided to prescribe medication immediately. The decision to use a medication may also reduce the exhaustion of the owner and considers previous attempts to correct this problem. Based on the clinical assessment the target neurotransmitter was serotonin. Dysfunction of the serotonergic system appears to be one of the reasons underlying ADHD, and especially in impulsivity (Oades, 2007). Fluoxetine, a selective serotonin reuptake inhibitor (SSRI), was selected at the dose of 3mg/kg PO sid (Mege *et al.*, 2003). Case reports of hyperactive dogs (Luescher, 1993; Piturru, 2014) often report the use of higher dosage of medication than the dose used in anxiety cases.

Epileptic seizures could be considered as a contraindication to the use of fluoxetine. Though fluoxetine has been deemed to reduce seizures threshold, recent studies found an opposite effect (Kanner, 2016; Wrzosek *et al.*, 2015). The Food and Drug Administration 2007 report[1] suggests that fluoxetine dose of 0.5 to 4.0 mg/kg sid can be used in canine patients.

Behavioural therapy started at the same time. The owner was instructed to reinforce calm behaviours, ignore signs of increasing arousal or confine the dog for short periods of time and maintain her composure near the dog. The owner was encouraged to play games with the dog in a calm manner and perform basic training with reward-based methods. Due to the dog's high arousal when presented with food, the owner was instructed to replace treats with praise or affection where possible. Finally, the owner was to be the one who initiates and directs play and interactions, not the dog.

Follow-up

The owner maintained contact via email between consultations. The dog showed no side effects of fluoxetine at the time of the follow-up. His behaviour has improved rapidly; within 1 week the owner noticed changes. The first sign of improvement was sleep duration; he started sleeping through the night without waking up.

However, during the day he was still highly active. Moreover, he initiated fewer undesirable interactions with the owner and the two other dogs in the house, and he was less destructive. Soon after that, the dog started to show signs and behaviours that he needs to soil, and his inappropriate elimination decreased. He did, however, continue to soil when he was highly aroused but to a lesser degree. The owner has limited the dog's access to water; however, at the same time, he seemed to drink less. Finally, the dog was slowly weaned off phenobarbital, and no seizures took place since.

Two months following the initial consultation, a follow-up consultation took place. At that time, the dog was much better. He slept well, and was calmer in familiar environments; however, he was still excited in new environments. The dog showed less arousal near his food but was highly aroused by human food. Similarly, he was drinking normal amounts of water. The dog urinated normally with minimal house soiling. However, he still voided urine when he was highly aroused or agitated. Although he was improving overall, there were times he was still difficult to handle.

During the follow-up consultation, he was calmer, could settle in the room, and only urinated a small amount when he left the consulting room and saw an unfamiliar dog. As opposed to the first visit, during the follow-up consultation, it was easier to perform a full physical exam. The dose of fluoxetine was raised to 3.5 mg/kg PO sid, to obtain further control. The amount was adapted to account for his weight gain.

Seven months following the initial consultations the dog continued to improve; his sleep and food intake were normal. He was less impulsive, more tolerant and less active. The dog still demonstrated high arousal when presented with human food and treats, and was moderately destructive at times. While he becomes excited easily, he was easier to control and more tolerant. The dog's water intake was normal, and he urinated appropriately most of the time with a few minor exceptions of urination when highly aroused or agitated.

Practical Application

Hypersensitivity–hyperactivity or ADHD is increasingly recognised in dogs although it is still controversial. Lack of satiety and hyposomnia, while described as characteristic of stage 2 HSHA, are poorly described in the literature (Pageat, 1998). Alimentary and sleep disorders have been discussed as part of ADHD in humans (Hvolby, 2015; Kaisari *et al.*, 2017).

This case report illustrated other symptoms that can be linked to stage 2 HSHA, including a putative connection between hyperexcitability and seizures. Psychogenic polydipsia is often an exclusion diagnosis. This case suggests HSHA as a possible underlying cause of psychogenic polydipsia.

While the recommended standard dose of fluoxetine is typically 1–2 mg/kg PO sid, this case demonstrated the use of a higher dose (up to 3.5 mg/kg PO sid). In cases of stage 2 HSHA, a higher dose should be considered to obtain adequate control.

Note

[1] FDA, 2007. NADA, 141-272. https://www.fda.gov/downloads/AnimalVeterinary/Products/ApprovedAnimalDrugProducts/FOIADrugSummaries/ucm062326.pdf

References

Campbell, W.E. (1973) Behavioral modification of hyperkinetic dogs. *Modern Veterinary Practice* 54, 49–52.

Hvolby, A. (2015) Associations of sleep disturbance with ADHD: implications for treatment. *ADHD Attention Deficit and Hyperactivity Disorders* 7, 1–18. DOI: 10.1007/s12402-014-0151-0.

Kaisari, P., Colin, T.D. and Higgs, S. (2017) Attention Deficit Hyperactivity Disorder (ADHD) and disordered eating behaviour: a systematic review and a framework for future research. *Clinical Psychology Review* 53, 109–121. DOI: 10.1016/j.cpr.2017.03.002.

Kanner, A.M. (2016) Most antidepressant drugs are safe for patients with epilepsy at therapeutic doses: a review of the evidence. *Epilepsy and Behavior*. Available at: http://dx.doi.org/10.1016/j.yebeh.2016.03.022.

Lit, L., Schweitzer, J.B., Iosif, A.M. and Oberbauer, A.M. (2010) Owner reports of attention, activity, and impulsivity in dogs: a replication study. *Behavioral and Brain Functions* 6, 1. DOI: 10.1186/1744-9081-6-1.

Luescher, U.A. (1993) Hyperkinesis in dogs: 6 case reports. *Canadian Veterinary Journal* 34, 368–370.

Marlois, N. (2001) Hyperactivity in dogs, a model for human pathology: discrepancies between different approaches. *In Proceedings of the Third International Congress on Veterinary Behavioural Medicine*, Vancouver, BC, 212–214.

Mege, C., Beaumont-Graff, E., Béata, C., Diaz, C., Habran, T., Marlois, N. and Muller, G. (2003) Pathologie comportementale du chien. Ed Masson, Paris.

Oades, R. (2007) The role of the serotonin system in ADHD: treatment implications. *Expert Review in Neurotherapeutics* 7, 1357–1374. doi.org/10.1586/14737175.7.10.1357.

Pageat, P. (1998) Pathologie du comportement du chien. Ed du Point Veterinaire, Maisons-Alfort.

Piturru, P. (2014) Methylphenidate use in dogs with attention deficit hyperactivity disorder (ADHD). *Tierärztliche Praxis Kleintiere* 2, 111–116.

Puurunen, J., Sulkama, S., Tiira, K., Araujo, C., Lehtonen, M., Hanhineva, K. and Lohi, H. (2016) A non-targeted metabolite profiling pilot study suggests that tryptophan and lipid metabolisms are linked with ADHD-like behaviours in dogs. *Behavioral and Brain Functions* 12, 27. doi.org/10.1186/s12993-016-0112-1.

Schoeman, J.P. (2008) Approach to Polyuria and Polydipsia in the dog. In-IVIS. Available at: http://www.ivis.org/proceedings/wsava/2008/lecture16/130.

Wrzosek, M., Plonek, M., Nicpon, J., Cizinauskas, S. and Pakozdy, A. (2015) Retrospective multicentre evaluation of the 'fly-catching syndrome' in 24 dogs: EEG, BAER, MRI, CSF findings and response to antiepileptic and antidepressant treatment. *Epilepsy & Behavior* 53, 184–189. Available at: http://dx.doi.org/10.1016/j.yebeh.2015.10.013.

Two Approaches to Managing Separation Anxiety

KERSTI SEKSEL*

Sydney Animal Behaviour Service, Seaforth, New South Wales, Australia

Conflict of interest: The author declares no conflict of interest.

Keywords: dog, separation anxiety

Introduction

Separation anxiety is the term used to describe the condition exhibited by dogs that are unable to cope without human company, often family members. These pets become extremely anxious and show distress behaviours such as vocalisation, destruction, house-soiling, inappetence, inactivity and even vomiting or diarrhoea in the total or virtual separation from people. The longer that these conditions are unrecognised and untreated, the more complex they appear to become and, therefore, potentially the harder to treat.

Diagnosis

Diagnosis is based on a complete behavioural history and thorough physical examination. It may involve complete blood work, dermatological and neurological work up as well as radiography to rule out contributing or concurrent medical factors.

Clinical signs

Dogs with this condition may follow people from room to room and begin to display signs of anxiety as soon as people prepare to leave. Some of these dogs crave

---

* sabs@sabs.com.au

©S. Denenberg 2017. *Proceedings of the 11th International Veterinary Behaviour Meeting* (ed. S. Denenberg)

a great deal of physical contact and attention from their owners, and they can be very demanding. During departures or separations, they may begin to salivate or pant profusely, vocalise, eliminate, refuse to eat, become destructive or become quiet and withdrawn. Not every dog exhibits all of the signs, but it appears that the more signs that are exhibited, the more difficult it will be to manage the case.

While these distress behaviours usually occur every time people leave, they may only be seen on selected departures, such as work-day departures, or when people leave again after coming home from work. Additionally, many dogs with noise phobias have concurrent separation anxiety and may exhibit marked destructiveness, house-soiling, salivation and vocalisation if this occurs when the owner is absent.

Management and follow-up

The aim of management is to teach the dog to tolerate being left alone. The owners are encouraged to keep a diary so that progress can be monitored. Concurrent or underlying other medical problems should be treated. All cases are followed up at regular intervals for several years.

One approach is to work through the 3 M approach plus 1 with almost equal emphasis on all 4 Ms. These are Behaviour Modification, Environmental Management, Medication and Monitoring. The 4 Ms approach means using behaviour modification with graduated departures, desensitisation and counter-conditioning as well as medications and environmental management.

The second method is first to focus on using medications to balance the levels of different neurotransmitters in the brain, most notably serotonin, and there is less emphasis on behaviour modification. Monitoring is always the key to determining progress.

Practical Applications

This paper discusses how the treatment of separation anxiety has evolved over the past two decades. Environmental management, behaviour modification and medication are still recommended for both methods. However, the approach has changed so that there is less effort involved for the owners and a similar (or better) response is achieved for the dog.

The Effects of a Nutritional Supplement (Solliquin) in Reducing Fear and Anxiety in a Laboratory Model of Thunder-induced Fear and Anxiety

GARY LANDSBERG[1]*, SCOTT HUGGINS[2], JULIE FISH[3] AND NORTON W. MILGRAM[1]

[1]CanCog Technologies, Toronto, Ontario, Canada; [2]Nutramax Laboratories Veterinary Science Inc, Lancaster, South Carolina, USA; [3]Vivocore Inc., Fergus, Ontario, Canada

Funding: This project was funded as contract research by Nutramax Laboratories to CanCog Technologies.

Conflict of interest: Scott Huggins is an employee of Nutramax Laboratories.

Keywords: dog, L-theanine, *Magnolia officinalis*, *Phellodendron amurense*, noise-induced anxiety

Introduction

The objective of this study was to evaluate the effectiveness of a nutritional supplement in reducing the anxiety response of Beagle dogs in a thunderstorm test model of noise-induced anxiety after one day of test article administration and after a double dose of the product following 7 days of administration. The supplement (Solliquin) is a nutraceutical intended for relief of clinical signs associated with fear, anxiety and stress, which contains a proprietary blend of L-theanine, extracts of *Magnolia officinalis* and *Phellodendron amurense*, as well as a whey protein concentrate previously found to have anti-anxiety effects. (Araujo *et al.*, 2010; DePorter *et al.*, 2012, 2016; Michelazzi *et al.*, 2015; Pike and Horwitz, 2015). The model has three components: a situational anxiety response evoked by the sound

---

* Corresponding author: garyl@cancog.com

of thunder, an anxiety response following exposure to a thunderstorm test, and a conditioned anxiety response to the test room associated with prior exposure to the thunderstorm (Araujo *et al.*, 2013).

Our initial description of the model targeted a reduction in activity in response to a sound stimulus (thunder recording) as the primary behavioural target (DePorter *et al.*, 2012; Araujo *et al.*, 2013). However, subsequent studies identified two types of responses to thunder sound exposure: first, dogs that exhibit both inactive (passive) responses (e.g. freeze, conflict-induced) and second, dogs that show active responses (flight or fight) (Dreschel and Granger, 2005; Mariti *et al.*, 2013; Mills *et al.*, 2013). Therefore, the data analysis for both anxiety measures and activity measures further account for these differences. We also use blood cortisol as measured 5 minutes after the completion of the thunder test as an anxiety index (Landsberg *et al.*, 2015).

Materials and Methods

Twenty-eight subjects, selected from a larger pool of thirty, were placed into two groups (placebo control and Solliquin) with equivalent global anxiety and cortisol responses to thunder. The study was run over a 15-day period. The first day was used to assess baseline activity in an open field arena. On subsequent tests, the animals were exposed in the same arena to a thunder recording protocol over two consecutive days as previously described published in which the dogs are exposed to 3 minutes' open field pre-thunder recording, 3 minutes of thunder recording and 3 minutes' post-thunder recording (Landsberg *et al.*, 2015). On the second and subsequent days of exposure, the first 3 minutes represents a conditioned pre-thunder anxiety. A baseline cortisol was collected 1 hour before testing and 5 minutes following the completion of the test. After allocation, dogs were treated for 7 days with either test article or placebo. On the second day of administration, a thunder test was performed 45 minutes following administration of test article or placebo. On the 7th day of administration, the dose amounts were doubled for each group, and an additional thunder test was performed. The activity assessment, conditioned anxiety and response to thunder response were assessed on both the 2nd day and 7th day of administration. The cortisol response was only assessed on the 7th day of treatment.

1. Activity data: All behavioural measurements were measured using Ethovision XT.
2. A trained technician scored the behaviour of the dogs using video files from the thunderstorm sessions. Scoring was based on an anxiety scale (Landsberg *et al.*, 2015), which provides for a measure of global anxiety, active anxiety and inactive (passive) anxiety. Each measure was scored on a 1–6-point visual analogue scale, which was scored blindly by the same trained observer for the entire study.
3. Cortisol: On the second baseline thunderstorm test, blood was collected 60 minutes prior to thunderstorm testing and 5 minutes following the thunderstorm session. Based on this initial analysis, subjects were classified as cortisol responders or non-cortisol responders. On the 7th day of treatment, blood was again collected from subjects classified as cortisol responders 60 minutes before the thunderstorm session and 5 minutes after the thunderstorm session.

Results

The study revealed a statistically significant decrease in distance travelled in response to thunder for the Solliquin treated animals after 1 day (p=0.0284). A marginally significant decline (p=0.0916), and a marginally significant group difference (p=0.0828) were found at day 7. The control group showed a non-significant decline at day 1 (p=0.5703) and day 7 (p=1.000) (Fig. 1). Solliquin treated dogs also showed a significant reduction (p=0.0435) in inactivity frequency after one day while the controls showed no effect (p=0.648). On day 7, inactivity duration also showed a marginally significant group difference (p=0.0828) with the Solliquin group increasing and the placebo group decreasing. Group mean comparisons were based on the Tukey's test.

A further analysis was set up to examine the effect of the test compound on the activity of subjects that showed a hyperactive response to noise. Five animals in each group were identified based on the animal's baseline response during the presentation of the thunder. There was a clear trend towards decreased activity in the subjects given the test compound, and the absence of a statistically significant difference could simply reflect low power due to the small number of animals in each group.

Based on baseline cortisol results, 22 dogs were identified as cortisol responders with 11 dogs allocated to each treatment group. Both groups showed lower post-thunder cortisol values during the treatment phase when compared to baseline, and there were no significant differences in mean change in post cortisol values between treatment and baseline phases between groups A and B (p=0.3374). In addition, an equal number of dogs in each group (8/11) showed an improved (decreased) cortisol response.

For the anxiety measures, there were no significant group differences.

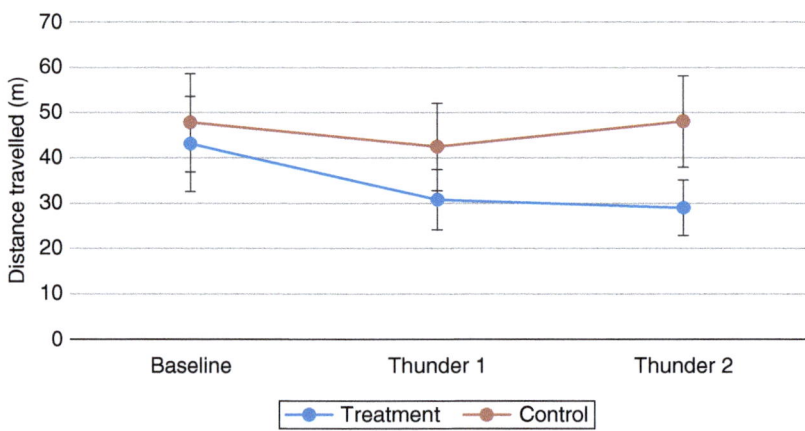

Fig. 1. Distance travelled as a function of group and test session during the presentation of thunder. Distance travelled decreased significantly from baseline (second baseline test) on the 2nd day of treatment and marginally significantly after day 7 with a marginally significant group difference. By comparison, the control group does not decrease significantly.

Discussion

The results support a calming effect of Solliquin treatment in reducing anxiety as demonstrated by a significant reduction in activity given that a hyperactive response is indicative of a strong anxiety response in these subjects. The effect of Solliquin was demonstrable on the second day of treatment where a significant reduction was seen in distance travelled and in inactivity frequency (the number of times an animal stops and starts during the testing) during thunder. On the 7th day of treatment, after a double dose of medication, during thunder there was a marginally significant group difference in distance travelled with the Solliquin group declining most, and in the time spent inactive with the Solliquin group declining and the control group increasing. Considering the clear trend towards decreased activity in the subjects identified as hyperactive that were given the test compound, further testing should consider a larger number of animals and the pre-selection of dogs that have a hyperactive response. Larger sample size would also provide a more homogeneous group for assessment of anxiety and cortisol measures.

References

Araujo, J.A., de Rivera, C., Ethier, J.A. *et al.* (2010) Anxitane® tablets reduce fear of human beings in a laboratory model of anxiety-related behavior. *Journal of Veterinary Behavior* 5, 268–275.

Araujo, J.A., de Rivera, C., Landsberg G.M. *et al.* (2013) Development and validation of a novel laboratory model of sound-induced fear induced anxiety in beagle dogs. *Journal of Veterinary Behavior* 8, 204–212.

DePorter, T.L., Landsberg, G.M., Araujo, J.L. *et al.* (2012) Harmonease® reduces noise-induced fear and anxiety in a laboratory canine model of thunderstorm simulation: a blinded and placebo-controlled study. *Journal of Veterinary Behavior* 7, 225–232.

DePorter, T.L., Bledsoe, D.L., Conley, J.R. *et al.* (2016) Case report series of clinical effectiveness and safety of Solliquin® for behavioral support in dogs and cats. *Proceedings of Veterinary Behavior Symposium*, pp. 27–28. San Antonio, Tx.

Dreschel, N.A. and Granger, D.A. (2005) Physiological and behavioral reactivity to stress in thunderstorm-phobic dogs and their caregivers. *Applied Animal Behavior Science* 95, 53–168.

Landsberg, G.M., Mougeot, I., Kelly, S. *et al.* (2015) Assessment of noise-induced fear and anxiety in dogs: modification by a novel fish hydrolysate supplemented diet. *Journal of Veterinary Behavior* 10, 391–398.

Mariti, C., Tancini, V., Gazzano, A., Calvo, P., Fatjo, J. and Bowen, J. (2013) Fear of thunder in dogs and owner related behavior; an Italian survey. In *Proceedings of the 9th International Veterinary Behavior Meeting*, p. 60. PSI, Lisbon, Portugal.

Michelazzi, M., Berteselli, G., Talamonti, Z. *et al.* (2015) Efficacy of l-theanine in treatment of noise phobias in dogs; preliminary results. *Veterinaria* 29, 1–7.

Mills, D., Braem Dube, M. and Zulch, H. (2013) Appendix B. The Lincoln Sound-sensitivity Scale. In: *Stress and Pheromonatherapy in Small Animal Clinical Behavior*. John Wiley and Sons, Chichester, UK, pp. 259–263.

Pike, A. and Horwitz, D.L. (2015) An open label prospective study of the use of l-theanine (Anxitane) in storm sensitive client owned dogs. *Journal of Veterinary Behavior* 10, 324–331.

Observing the Results of Reducing the Stress of Dogs During Training by the Help of Dog-appeasing Pheromone on Learning and Problem-solving Behaviours

ETKIN SAFAK* AND NESRIN SULU

Department of Physiology, Faculty of Veterinary Medicine, Ankara University, Ankara, Turkey

Funding: Ankara University Scientific Research Projects Coordination Unit have supported this research. Project Number: 15L0239002, 2015. This research was confirmed with ethical approval (Protocol No: 2014-16-96) by Ethical Committee of Ankara University.

Conflict of interest: The authors declare no conflict of interest.

Keywords: dog training, Dog Appeasing Pheromone, problem solving, working dogs, stress

Introduction

Working dogs are subjected to a strict training in order to be able to fulfil their future duties in real life. This study aims to investigate the effects of the Dog Appeasing Pheromone (DAP; Adaptil®) on reducing stress during training as well as on learning and problem-solving abilities of dogs.

Materials and Methods

The present study was conducted at the Ministry of Customs and Trade, General Directorate of Customs Enforcement, Dog Training Center in Turkey. Thirty-three

* Corresponding author: esafak@ankara.edu.tr

Belgian Malinois dogs aged between 6 months and 3 years old were used. These dogs were divided into three groups (G1, G2, G3). G1 (n=11; without DAP) and G2 (n=11; with DAP) underwent basic obedience training for 7 days. At the end of the training both groups performed a basic learning test and a problem-solving test (detour test). G3 (n=11) was brought to the training area and spent the same time as in the basic obedience for 7 days without using pheromone collars and/or training. Cortisol level, heart rate per minute, body temperature and behavioural evaluation were used as stress parameters. Results were analysed using SPSS v18.0.

Results

All physiological parameters, and behavioural indicators such as lowered head, tail, ear and body postures were statistically lower in G2 in comparison to those in G1 (Mann–Whitney U test; $p < 0.01$). Furthermore, G2 performed better in problem solving and learning tests (Mann–Whitney U test; $p < 0.05$).

Conclusion

These results suggest that Adaptil is a promising tool to reduce stress and increase performance during training.

Evaluation of the Association Between Attendance at Veterinary Hospital-based Puppy Socialisation Classes and Long-term Retention in the Home

RICK SCHULKEY[1] AND THERESA DEPORTER[2]*

[1]Madison Veterinary Hospital, Madison Heights, Michigan, USA; [2]Oakland Veterinary Referral Services, Bloomfield Hills, Michigan, USA

Conflict of interest: The authors declare no conflict of interest.

Keywords: Canine, socialisation, veterinary hospital, retention

Introduction

Puppies bring new clients to the veterinary hospital, but many dogs are relinquished to a new home, a shelter or euthanised before their second birthday (Lund *et al.*, 1999; New *et al.*, 2000; Clancy and Rowan, 2003).

Materials and Methods

Following a retrospective review of files of Madison Veterinary Hospital, in 2009, owners were contacted. Approximately two-thirds of the puppies were no longer in their original homes by their second birthday. Behaviour advice given during puppy visits were transformed from dominance and punishment philosophy to a positive reinforcement programme. Low-stress handling techniques were initiated (Yin, 2009). Puppies between 7 and 12 weeks of age were offered the opportunity to attend five weekly in-hospital puppy socialisation classes called 'Puppy Kindergarten' (PK). Between January 2010 and July 2014, 519 puppies participated in PK conducted by hospital employees. Educational materials were

* Corresponding author: Theresadax@aol.com

presented including medical and behaviour recommendations. Canine body language was discussed, and positive reinforcement training was demonstrated. Puppies were familiarised with all aspects of the hospital and the veterinary staff.

Results

Out of 519 dogs, 507 were located, and 491 were still with their original family. Owner interviews identified dogs that were not in the original home: ten were surrendered due to owner-related housing changes, one was euthanised due to behaviour, and six were deceased due to medical conditions.

Conclusion

In this hospital, the long-term retention rate for puppies was improved from approximately 33% to 94.4% remaining in the home by incorporating 'Puppy Kindergarten' classes within the veterinary hospital (Scarlett *et al.*, 2002; Duxbury *et al.*, 2003).

References

Clancy, E. and Rowan, A. (2003) Companion animal demographics in the United States: A historical perspective. In: Salem, D.J. and Rowan, A.N. (eds) *The State of the Animals II*. Humane Society Press, Washington, DC, pp. 9–26.

Duxbury, M.M., Jackson, J.A., Line, S.W. and Anderson, R.K. (2003) Evaluation of association between retention in the home and attendance at puppy socialization classes. *Journal of the American Veterinary Medical Association* 223, 61–66. DOI: 10.2460/javma.2003.223.61.

Lund, E.M., Armstrong, P.J. and Kirk, C.A. (1999) Health status and population characteristics of dogs and cats examined at private veterinary practices in the United States. *Journal of the American Veterinary Medical Association* 214, 1336–1341.

New, J.C., Salman, M.D., King, M., Scarlett, J.M., Kass, P.H. and Hutchison, J.M. (2000) Characteristics of shelter-relinquished animals and their owners compared with animals and their owners in U.S. pet-owning households. *Journal of Applied Animal Welfare Science* 3, 179–201. DOI: 10.1207/S15327604JAWS0303_1.

Scarlett, J.M., Salman, M.D., New, J.G. and Kass, P.H. (2002) Exploring the bond: the role of veterinary practitioners in reducing dog and cat relinquishments and euthanasias. *Journal of the American Veterinary Medical Association* 220, 306–311.

Yin, S. (2009) *Low Stress Handling, Restraint and Behaviour Modification of Dogs & Cats*. CattleDog Publishing, Davis, California.

Effects of Olfactory Stimulation with Essential Oils in Animals: A Review

Anouck Haverbeke[1]*, Stijn Schoelynk[2], Adinda Sannen[2], Hilde Vervaecke[2], Chiara Mariti[3], Jara Gutiérrez Rufo[3], Angelo Gazzano[3] and Stefania Uccheddu[1]

[1]Vet Ethology, Overijse, Belgium; [2]Ethology & Animal Welfare, Agro- & Biotechnology, ODISEE University College, Sint-Niklaas, Belgium; [3]EtoVet, Dip. Scienze Veterinarie, Università di Pisa, Pisa, Italy

Conflict of interest: The authors declare no conflict of interest.

Keywords: animals, emotions, essential oils, olfactory stimulation, welfare

Introduction

The influence of essential oils on emotions has been widely described for humans (Roberts and Williams, 1992), zoo animals (Schuett and Frase, 2001; Pearson, 2002; Wells and Egli, 2004), cats (Ellis, 2009), dogs (Wells, 2004; Graham *et al.*, 2005) and horses (Ferguson *et al.*, 2013). Specific scents trigger memories and emotions by activating the amygdala and the hippocampus. Inhalation of essential oils may lead to a profound psychological and physiological effect via the same pathway (Cahill *et al.*, 1995). Citrus or sweet orange essences have been used with positive results in rats with anxiety symptoms (Faturi *et al.*, 2010). The inhalation of lavender led to a significant decrease in the heart rate in horses that had been exposed to acute stress (Ferguson *et al.*, 2013). Lavender and chamomile had a similar effect on dogs (Graham *et al.*, 2005).

Discussion

The scientific literature demonstrates that essential oils may have beneficial effects on both the behaviour and welfare of non-human animals. This suggests that

---

* Corresponding author: anouck.haverbeke@vetethology.be

essential oils may have a potential practical application in behavioural therapy and welfare improvement in animals. More research should be conducted to further investigate the potential of olfactory stimulation in the prevention of behavioural problems (such as at animal shelters, at the owners' home after adoption, or at the veterinary clinic) or for the treatment of behavioural problems among animals.

References

Cahill, L., Babinsky, R., Markowitsch, H.J. and McGaugh, J.L. (1995) The amygdala and emotional memory. *Nature* 377, 295–296.

Ellis, S.L. (2009) Environmental enrichment: practical strategies for improving feline welfare. *Journal of Feline Medicine & Surgery* 11(11), 901–912.

Faturi, C.B., Leite, J.R., Alves, P.B., Canton, A.C. and Teixeira-Silva, F. (2010) Anxiolytic-like effect of sweet orange aroma in Wistar rats. *Progress in Neuro-Psychopharmacology & Biological Psychiatry* 34, 605–609.

Ferguson, C.E., Kleinman, H.F. and Browning J. (2013) Effect of lavender aromatherapy on acute-stressed horses. *Journal of Equine Veterinary Science* 33(1), 67–69.

Graham, L., Wells, D.L. and Hepper, P.G. (2005) The influence of olfactory stimulation on the behaviour of dogs housed in a rescue shelter. *Applied Animal Behavior Science* 91(1), 143–153.

Pearson, J. (2002) On a roll: novel objects and scent enrichment for Asiatic lions. *Shape Enrichment* 11(7).

Roberts, A. and Williams, J.M.G. (1992) The effect of olfactory stimulation on fluency, vividness of imagery and associated mood – a preliminary study. *British Journal of Medical Psychology* 65, 97–199.

Schuett, E.B. and Frase, B.A. (2001) Making scents: using the olfactory senses for lion enrichment. *The shape of Enrichment* 10(3), 1–3.

Wells, D.L. (2004) A review of environmental enrichment for kennelled dogs (*Canis familiaris*). *Applied Animal Behavior Science* 85(3), 307–317.

Wells, D.L. and Egli, J.M. (2004) The influence of olfactory enrichment on the behaviour of captive black-footed cats (*Felis nigripes*). *Applied Animal Behavior Science* 85(1), 107–119.

Olfactory Enrichment in Dogs: Possible New Applications

Stefania Uccheddu[1]*, Stijn Schoelynk[2], Adinda Sannen[2], Hilde Vervaecke[2], Heidi Arnouts[2], Jara Gutiérrez Rufo[3], Chiara Mariti[3], Angelo Gazzano[3] and Anouck Haverbeke[1]

[1]Vet Ethology, Overijse, Belgium; [2]Ethology & Animal Welfare, Agro- & Biotechnology, ODISEE University College, Sint-Niklaas, Belgium; [3]EtoVet, Dip. Scienze Veterinarie, Università di Pisa, Pisa, Italy

Funding: This study was funded by the American Holistic Veterinary Medical Foundation.

Conflict of interest: The authors declare no conflict of interest.

Keywords: behaviour, essential oils, olfactory enrichment, shelter dogs, welfare

Introduction

Environmental enrichment improves the overall well-being of animals living in captivity. In stressful conditions, enrichment may increase the animal's ability to cope, thereby improving its welfare. Several enrichment programmes have been developed in shelters to meet animal needs. Although current literature shows the effectiveness of olfactory enrichment in welfare improvement in dogs, cats and horses (Wells, 2004, 2009; Graham et al., 2005; Ellis, 2009; Ellis and Wells, 2010; Ferguson et al., 2013) only few such programmes exist. Certain odours resulted in behaviour indicative of increased relaxation in shelter dogs (Graham et al., 2005). Captive cats exposed to experimental odours (Wells, 2004) showed increased activity, reduced sedentary behaviours, and increased exploration behaviour, all changes that could be interpreted as indicators of increased welfare. The aim of this study was to investigate whether olfactory stimulation could have a beneficial effect.

---

* Corresponding author: uccheddus@gmail.com

Material and Methods

A group of shelter dogs (n=110) underwent behavioural observation, a cognitive bias test, and measurement of their salivary cortisol. The effect of nine essentials oils was evaluated. The oils were diffused separately and in combination.

Results

From our initial observation, dogs appeared more relaxed when essential oils were administered. Further results and statistical analyses are pending.

Discussion and Conclusion

Olfactory stimulation appears to be a promising tool in the prevention or reduction of anxiety in shelter dogs. It can easily be implemented in shelters as a common practice, together with environmental and social types of enrichment. Further research is needed to assess its long-term effects on dog welfare.

References

Ellis, S.L.H. (2009) Environmental enrichment: practical strategies for improving feline welfare. *Journal of Feline Medicine and Surgery* 11, 901–912.

Ellis, L.H.S. and Wells, D.L. (2010) The influence of olfactory stimulation on the behaviour of cats housed in a rescue shelter. *Applied Animal Behavior Science* 123, 56–62.

Ferguson, C.E., Kleinman, H.F. and Browning, J. (2013) Effect of lavender aromatherapy on acute-stressed horses. *Journal of Equine Veterinary Science* 33, 67–69.

Graham, L., Wells, D.L. and Hepper, P.G. (2005) The influence of olfactory stimulation on the behaviour of dogs housed in a rescue shelter. *Applied Animal Behavior Science* 91, 143–153.

Wells, D.L. (2004) A review of environmental enrichment for kennelled dogs (canis familiaris). *Applied Animal Behavior Science* 85, 307–317.

Wells, D.L. (2009) Sensory stimulation as environmental enrichment for captive animals: a review. *Applied Animal Behavior Science* 118, 1–11.

Hair Cortisol in Cats as a Measure of Chronic Stress: A Pilot Study Under Controlled Conditions

Marta Amat[1], Ana García[2], Camino García-Morato[1]*, Déborah Temple[1], Susana Le Brech[1], Tomàs Camps[1] and Xavier Manteca[1]

[1]*School of Veterinary Medicine (Autonomous University of Barcelona), Bellaterra (Cerdanyola del Vallès), Spain;* [2]*ENVIGO CRS, S.A.U. Centro Industrial Santiga, Barcelona, Spain*

Conflict of interest: The authors declare no conflict of interest.

Keywords: cats, chronic stress, hair cortisol

Introduction

Stress-related problems are a major welfare issue in cats. Hair cortisol has been proposed as a measure to assess the long-term activity of the hypothalamic–pituitary–adrenal axis (HPA). In this study, we measured hair cortisol in cats kept under controlled conditions.

Material and Methods

Thirty-six domestic shorthair cats were included in this study. Cats had been housed in groups separated by gender to reduce aggressive behaviour before commencing the study. The study was divided into two phases. In Phase I, which lasted for 3 months, cats were individually housed in cages, whereas in Phase II, cats were housed in groups, separated by sex, in two different rooms. A hair sample was collected from each animal at the beginning and at the end of Phase I

---

* Corresponding author: Camino.GarcíaMorato@uab.cat

(samples A and B respectively) and 3 months after the start of Phase II (sample C). An enzyme-linked immunoassay (ELISA) test was used to determine hair cortisol levels. Data were analysed using the MIXED procedure of SAS with Tuckey adjustment and including the variable cat as a repeated measure.

Results

Cortisol levels were significantly higher in sample B (9.4 ± 0.33 pg/mg) than in samples A (7.2 ± 0.30 pg/mg) and C (7.1 ± 0.31 pg/mg), indicating that individually housed cats had higher levels of hair cortisol than group-housed cats.

Conclusions

The results of this study suggest that hair cortisol levels may be a valid measure of chronic stress in cats. Moving from group to individual housing may be stressful for cats.

Cortisol Content in Hair Measured by Liquid Chromatography–Tandem Mass Spectrometry: A Non-invasive Marker of Chronic Stress in Companion Animals

ISABELLE MOUGEOT[1]*, JULIE FISH[2], INES DE LANNOY[3], ANJA GRUJIC[3] AND GARY LANDSBERG[1]

[1]CanCog Technologies, Toronto, Ontario, Canada; [2]Vivocore, Fergus, Ontario, Canada; [3]Intervivo Solutions, Toronto, Ontario, Canada

Conflict of interest: The authors declare no conflict of interest.

Keywords: chronic stress, dogs, cortisol hair content, liquid chromatography–tandem mass spectrometry

Introduction

Serum and saliva cortisol levels are often reliable measures of the acute stress response in companion animals. However, there is limited knowledge regarding cortisol hair content as a measure of chronic stress in pets (Park *et al.*, 2016). This study investigated baseline hair cortisol levels from healthy dogs, atopic dogs, healthy cats, or cats suffering from feline idiopathic cystitis (FIC). Hair cortisol measures could help to better monitor stress-related chronic illnesses such as atopy in dogs and FIC in cats. Measuring the cortisol hair content using liquid chromatography–tandem mass spectrometry (LC-MS/MS) technique could offer such a non-invasive method.

* Corresponding author: isabellem@cancog.com

Material and Methods

Hair was collected from 23 atopic dogs, 13 healthy dogs, 8 cats with FIC and 25 healthy cats. For each animal, approximatively 150 mg of hair was collected from three body areas. LC-MS/MS was performed on samples to quantify hair cortisol concentration.

Statistical analyses were run to compare cortisol levels from healthy and atopic dogs (Mann–Whitney U test), as well as from healthy and FIC cats (Student t-test).

Results

The baseline hair cortisol levels were significantly higher ($p=0.0321$) in atopic (med: 4.49, min: 1.49, max: 5.00) than in healthy (med: 2.63, min: 1.34, max: 4.20) dogs. The baseline hair cortisol levels were not significantly different in healthy (mean: 2.3970, SD: 0.5225) and FIC cats (mean: 2.7386, SD: 0.8078).

Conclusion

Hair cortisol content evaluated by LC-MS/MS appears to reflect the dog's chronic stress when suffering from atopy. Unanticipated findings in cats affected by FIC suggest that this cannot be established yet for feline subjects.

References

Park, S.H., Kim, S.A., Shin, N.S. and Hwang, C.Y. (2016) Elevated cortisol content in dog hair with atopic dermatitis. *Japanese Journal of Veterinary Research* 64(2), 123–129.

Association Between Catecholaminergic Genes and Impulsivity in Dogs

Claire Colsoulle[1], Jean-Yves Matroule[2], Jérôme Copine[2], Benoit Bihin[2], Eric Depiereux[2] and Claire Diederich[2]*

[1]Université de Louvain, Belgium; [2]University of Namur, Belgium

Conflict of interest: The authors declare no conflict of interest.

Keywords: genetic markers, canine, dopamine transporter, personality trait, tyrosine hydroxylase

Introduction

The catecholaminergic system-related genes are known to control for neurotransmitters influencing behaviours. Dogs' impulsivity is now considered as a personality trait involving higher general activity throughout contexts. This study aimed at identifying possible associations between a selection of catecholaminergic genes (*SLC6A3, TH, COMT, ADRB1*[1]) and dogs' impulsivity scores.

Materials and Methods

Eighty-three pet dogs (33 Border Collies, 15 German Shepherds, 29 Labradors and 6 Golden Retrievers) were scored for impulsivity with three behavioural tests (spontaneous activity on a leash, the approach of an unknown person, and oral manipulation) (Wan *et al.*, 2013). Additionally, owners completed two questionnaires describing dogs and owners, their living places, dogs' personality, dogs' reactivity (including impulsivity), and dog–owner relationship (Vas *et al.*, 2007; Wright *et al.*, 2011). Seventeen dogs were re-tested 8 months later. DNA samples were collected (buccal scrape) and the genotyping of the catecholaminergic

---

* Corresponding author: claire.diederich@unamur.be

genes was processed (*SLC6A3* and *TH* genes, classical PCR and electrophoresis, identification of three genotypes per gene; *COMT*, high resolution melting (HRM) method, two genotypes identified; *ADRB1* gene's amplification had been unsuccessful). Behavioural data were summarised with PCA and statistics were computed according to data types (Pearson, ANOVA, significant level p<0.05).

Results

Test–retest results were highly correlated (behavioural tests: r=0.66, DIAS: r=0.68, Dog-ADHD-RS: r=0.76). Within impulsivity scores, correlations were low (r=0.16 (DIAS/Dog-ADHD-RS correlation), r=0.35 (behavioural test/Dog-ADHA-RS)), to moderate (r=0.60 (DIAS/Dog-ADHD-RS)). Results showed that genotypic frequencies differed according to breeds (*SLC6A3* gene, p=0.0004; *TH* gene, p=0.0002, *COMT* gene: p=0.0095). No significant correlation was observed between each impulsivity score (behaviour or questionnaire) and each of the three genes.

Conclusion

This study did not succeed in identifying a link between catecholaminergic genes and impulsivity in dogs. It is suggested that behavioural phenotype should be improved (low scores' correlations) and to join functional study (effect of the mutation under study on protein synthesis and function) to the analyses.

Note

[1] *SLC6A3*: Dopamine Transporter or DAT; *TH*: Tyrosine hydroxylase; *COMT*: Catechol-O-Methyl Transferase; *ADRB1* Beta-ADrenergic 1 Receptor.

References

Vas, J., Topal, J., Pech, E. and Miklosi, A. (2007) Measuring attention deficit and activity in dogs: a new application and validation of a human ADHD questionnaire. *Applied Animal Behavior Science* 103(1–2),105–117.

Wan, M., Hejjas, K., Ronai, Z., Elek, Z., Sasvari-Szekely, M., Champagne, F.A., Miklosi, A. and Kubinyi, E. (2013) DRD4 and TH gene polymorphisms are associated with activity, impulsivity and inattention in Siberian Husky dogs. *Animal Genetics* 44(6) 717–727.

Wright, H.F., Mills, D.S. and Pollux, P.M.J. (2011) Development and validation of psychometric tool for assessing impulsivity in the domestic dog (*Canis familiaris*). *International Journal of Comparative Psychology* 24(2), 210–225.

Do Assistance Dogs Show Work Overload? Canine Blood Prolactin as a Clinical Parameter to Detect Chronic Stress-related Response

Manuel Mengoli[1]*, Tiago Mendonça[1], Jessica Lee Oliva[1], Cécile Bienboire-Frosini[1], Camille Chabaud[1], Elisa Codecasa[1], Muriel Jochem[2], Alessandro Cozzi[1] and Patrick Pageat[1]

[1]IRSEA, Research Institute in Semiochemistry and Applied Ethology, Quartier Salignan, France; [2]Fondation Frédérique Gaillanne - MIRA Europe, L'Isle-sur-la-Sorgue, France

Conflict of interest: The authors declare no conflict of interest.

Keywords: assistance dog, behavioural therapy, prolactin, stress-related response

Introduction

Owners and trainers of assistance dogs must be able to manage the dog's stress levels efficiently. These dogs are exposed to various situations every day and are expected to perform complicated tasks. It is a fundamental aspect that the dog's welfare is maintained and monitored. The aim of this study was to investigate the use of blood prolactin level as an indicator of a chronic stress response.

Material and Methods

Twenty-one assistance dogs (AD) and 25 pet dogs (PD) were compared using a behavioural assessment (Table 1). Females involved were neutered or in anoestrus. Pets' routine had a lower impact on their emotional state, while all AD were in

* Corresponding author: m.mengoli@group-irsea.com

Table 1. Assistance (AD) and pet dogs (PD) population, regarding gender and sex status.

	AD	PD
Entire male	0	4
Neutered male	9	13
Entire female	1	2
Neutered female	11	6

continuous training (at the school, during the week; with the forecaster families, the weekend). Blood samples were collected from all dogs and prolactin level was measured as a biological parameter. Results were analysed using Wilcoxon test.

Results

Assistance dogs had a statistically significant higher mean prolactin blood level (14.73±3.81 ηg/ml) than PD (6.05±1.80 ηg/ml). A physiological range under 15 ng/ml is considered normal. Two AD presented persistent hyperprolactinemia (51 ηg/ml and 33 ηg/ml). The same two dogs exhibit lower work performance due to restlessness, displacement activities, and increased vigilance during work. These dogs were suspended as AD to perform other clinical analyses and possible behavioural therapies.

Conclusion

These results show the necessity to find new clinical approaches to efficiently detect emotional disturbances in assistance dogs and to increase their welfare. Canine blood prolactin can become an interesting clinical parameter to detect a sustained stress-related response in assistance dogs.

Efficacy of a Therapeutic Diet on Dogs with Signs of Cognitive Dysfunction Syndrome

Gary Landsberg[1]*, Yuanlong (Gary) Pan[2], Isabelle Mougeot[1], Stephanie Kelly[3], Hui Xu[2], Sandeep Bhatnagar[2] and Norton W. Milgram[1]

[1]CanCog Technologies, Toronto, Ontario, Canada; [2]Nestlé Purina Research, St. Louis, Missouri, USA; [3]Vivocore Incorporated, Fergus, Ontario, Canada

Funding: This project was funded as contract research by Nestlé Purina Research to CanCog Technologies.

Conflict of interest: Yuanlong Pan, Hui Xu and Sandeep Bhatnagar are employees of Nestlé Purina Research.

Keywords: antioxidants, arginine, cognitive dysfunction syndrome, medium chain triglycerides, brain nutrient blend

Introduction

Previous studies have shown that diets supplemented with medium chain triglyceride oil (MCT) or brain protection blend nutrient blend (BPB)[1] enhance learning, memory and problem-solving ability in senior pets (Pan *et al.* 2010, 2013). The objective of this clinical study was to evaluate the effects of a diet supplemented with MCT and BPB on pet dogs with cognitive dysfunction syndrome (CDS).

Materials and Methods

Participating veterinary clinics screened senior dogs for signs of CDS as determined by a Senior Canine Behaviour and Health Questionnaire, which included six categories (disorientation, social interactions, sleep–wake cycle, loss of house

---

* Corresponding author: garyl@cancog.com

©S. Denenberg 2017. *Proceedings of the 11th International Veterinary Behaviour Meeting* (ed. S. Denenberg)

training, anxiety and altered activity). Out of 102 dogs, after ruling out potential medical causes, 29 were enrolled and randomised into one of three groups with either 9%, 6.5% or 0% MCT. Fisher least significant difference test was used to analyse the findings.

Results

All six categories of CDS were significantly ($p < 0.05$) improved in the dogs given the 6.5% MCT diet at the end of the 90-day study. The control diet did not significantly improve disorientation and social interaction. At 30 days five of six categories were significantly ($p < 0.05$) improved in the 6.5% MCT group compared to three of six categories in the placebo group. The 9% MCT diet produced significant improvement in only those dogs whose owners had positive feedback on the diet, suggesting possible palatability issues.

Discussion

These findings support the use of the 6.5% MCT and BPB supplemented diet[2] for the treatment of senior dogs with clinical signs consistent with CDS.

Notes

[1] BPB contains docosahexanoic acid (DHA), eicosapentanoic acid (EPA), B vitamins, antioxidants and arginine.
[2] Purina ProPlan® Veterinary Diet NC Neurocare.

References

Pan, Y.L., Larson, B., Araujo, J.A. *et al.* (2010) Dietary supplementation with medium-chain TAG has long-lasting cognition-enhancing effects in aged dogs. *British Journal of Nutrition* 103, 1746–1754.
Pan, Y.L., Araujo, J.A., Burrows, J. *et al.* (2013) Cognitive enhancement in middle-aged and old cats with dietary supplementation with a nutrient blend containing fish oil, B vitamins, antioxidants and arginine. *British Journal of Nutrition* 110, 40–49.

Social Behaviour and Bonding in Aged Dogs: A Multimodal Assessment Approach

Patricia Darder[1]*, Anna Salas[2], Elena García[1], Jaume Ferrer Lalanza[3], Jaume Fatjó[2,3], Celina Torre[2] and Jon Bowen[3]

[1]Ethogroup, Barcelona, Spain; [2]R&D Department, Affinity Petcare, Barcelona, Spain; [3]Ethometrix Ltd, Hove, UK

Funding: The present study has been developed with the financial support of Affinity Pet Care, S.A.

Conflict of interest: Anna Salas, Jaume Fatjó and Celina Torre are associated with Affinity Petcare.

Keywords: aged dogs, attachment, social behaviour, cognitive decline

Introduction

Most studies on ageing are usually focused on one dimension of behaviour. A study was designed to explore changes in social behaviour and bonding in a group of 13 aged dogs undergoing environmental enrichment and receiving a supplemented diet.

Materials and Methods

Social behaviour and bonding was evaluated through the strange-situation test (SST). An accelerometer was used to assess behaviour before, during and after the SST. Visual-spatial memory was assessed through a task based on a discriminant instrumental learning paradigm. Blood was tested for complete blood count (CBC), biochemical profile and oxidative markers. All measurements were taken

* Corresponding author: etologia.darder@gmail.com

©S. Denenberg 2017. *Proceedings of the 11th International Veterinary Behaviour Meeting* (ed. S. Denenberg)

at three time points during 5 months. All dogs received the same diet supplemented with antioxidants and enrichment. Statistical analysis included analysis of variance (ANOVA), principal component analysis (PCA) and partial least squares-discriminant analysis (PLS-DA).

Results

Memory test performance significantly declined throughout the study. A significant reduction in some oxidative markers was also observed throughout the study. Poorer performing dogs had significantly lower levels of superoxide dismutase. In the SST, by the end of the study dogs showed significantly less exploratory behaviour, interacted more with the caregiver and modified their pattern of interaction with the stranger. Daytime activity became less random and night activity decreased, indicating a more balanced behaviour pattern.

Discussion and Conclusions

The ageing process and interventions to counteract mental decline seemed to differentially impact various cognitive processes. Changes in social behaviour and bonding could help aged dogs to better cope with environmental challenges. Our results stress the value of looking at different biological and behavioural dimensions to explore the consequences of ageing on cognition and adaptation.

Bibliography

Mongillo, P., Pitteri, E., Carnier, P., Gabai, G., Adamelli, S. and Marinelli, L. (2013) Does the attachment system towards owners change in aged dogs? *Physiology & Behavior* 120, 4–69.

Szabó, D., Gee, N.R. and Miklósi, A. (2016) Natural or pathologic? Discrepancies in the study of behavioral and cognitive signs in aging family dogs. *Journal of Veterinary Behavior: Clinical Applications and Research* 11, 86–98.

Olfactory Stem Cell Therapy in Canine Age-related Disorders Treatment: A Controlled Study

Antoine D. Veron[1,2]*, Manuel Mengoli[3], Cécile Bienboire-Frosini[1], Violaine Mechin[1], Elisa Codecasa[1], Coralie Scifo[1], Romain Pageat[3], Gary Landsberg[4], Alessandro Cozzi[1], Joseph Araujo[5], Pietro Asproni[1] and Patrick Pageat[1,3]

[1]IRSEA, Research Institute in Semiochemistry and Applied Ethology, Apt, France; [2]Aix Marseille Univ., CNRS, NICN, Marseille, France; [3]CECBA, Clinical Ethology and Animal Welfare Centre, Apt, France; [4]CanCog Technologies, Toronto, Ontario, Canada; [5]InterVivo Solutions, Inc., Fergus, Ontario, Canada

Funding: Antoine D. Veron is the recipient of a Convention Industrielle de Formation par la Recherche (CIFRE) fellowship from the Association Nationale de la Recherche et de la Technologie (ANRT).

Conflict of interest: The authors declare no conflict of interest.

Keywords: olfactory, vecto-mesenchymal stem cells, ageing, dog, cellular therapy

Introduction

Ageing is an unavoidable process that may become pathological. Thus, very commonly age-related disorders affect dogs, leading to neuronal losses impacting cognitive and emotional capabilities (Landsberg *et al.*, 2012). At this time, no therapies are completely effective, and this often leads to euthanasia. In this context, stem cell-based therapies emerged as a conceivable therapy to face brain degeneration.

Mesenchymal stem cells (MSC) are recognised for their ability to protect the brain against injuries through secretion of various molecules that can enhance

---

* Corresponding author: a.veron@group-irsea.com; antoine.veron@univ-amu.fr.

neurogenesis (Maltman *et al.*, 2011) and limit apoptosis (Uccelli *et al.*, 2011). Among the type of MSC for use in regenerative therapies, the olfactory ecto-mesenchymal stem cells (OE-MSCs), residing in the olfactory mucosa, stand out as a promising option (Murrell *et al.*, 2005; Delorme *et al.*, 2010). Indeed, this peripheral and permanently self-renewing nerve tissue contains potent and highly proliferative stem cells. They can notably differentiate into cells of the neural lineage (Nivet *et al.*, 2011) and secrete neurotrophic and immunomodulatory factors (Di Trapani *et al.*, 2013; Ould-Yahoui *et al.*, 2013).

These cells present the advantage of being easily accessible in dogs and other species (Veron *et al.*, 2017; personal data) and can be quickly and easily multiplied in large enough numbers for cell transplantations (Shafiee *et al.*, 2011). The effect of their transplantation has also been evaluated in various rodent models of tissue damage such as Parkinson's disease (Murrell *et al.*, 2008), global cerebral ischaemia (Veron *et al.*, 2017; personal data) and in a model of hippocampal lesions (Nivet *et al.*, 2011). Previously, we also described that OE-MSCs transplanted in the fourth ventricle of two dogs could reduce some age-related disorder symptoms (Pageat *et al.*, 2014; Veron *et al.*, 2014; Mengoli *et al.*, 2016). Thus, we hypothesised that these mesenchymal stem cells from a nerve tissue may represent a well-suited tool for the treatment of age-related disorders in dogs.

The aim of this study was to evaluate whether OE-MSCs grafting in aged dogs would restore cognitive abilities and reduced emotional issues.

Material and Methods

For this study, nine female Beagles (9 years old) were included after allowance by IRSEA's Ethical Committee (C2EA125). Dogs followed a standard pre-training protocol before the beginning of the study (Tapp *et al.*, 2003; Snigdha *et al.*, 2012).

Two biopsies of olfactory mucosa measuring 1mm^2 were collected on anaesthetised and intubated dogs. Tissues were placed in culture medium and transported at 4°C until dilaceration and seeding in culture wells. Cells were cultured as previously described for rat cells (Stamegna *et al.*, 2014). Before grafting, cells were characterised for stem cell markers and compared to human OE-MSCs (Delorme *et al.*, 2010).

During growing of OE-MSCs cognitive and emotional abilities of dogs were evaluated using the Age-Related Cognitive and Affective Disorders/'Evaluation du vieillissement émotionnel et cognitif' scale (ARCAD/EVEC) (Colle *et al.*, 2000). In addition, dogs were subjected to different tests using the Toronto General Testing Apparatus (TGTA) (Araujo *et al.*, 2008): (i) an associative discrimination task and an attention task with various number of distractors were used for learning capacity assessment; (ii) a reversal of learned rules was performed to evaluate emotional instability; and (iii) a delayed non-matching to place (DNMP) test was used to assess long-term memory (Table 1).

Following 4 weeks of baseline evaluation (T0), dogs were randomly allocated to two groups and grafted with 200 million of their own OE-MSCs (n=4) or culture medium (n=5). To proceed with injection of stem cells in the fourth ventricle,

Table 1. Experimental design of TGTA based tests.

Week	Test	Trials per days/week
1	Discrimination	20/100
2	Attention	12/60
3	Reversal	20/100
4	Variable DNMP (delay: 5, 55, 105 sec)	18/90

dogs were anaesthetised and an endotracheal tube was placed for isoflurane delivery. Following 4 weeks' rest, the same four tests were repeated for 4 weeks (T1) followed by 4 weeks' rest and so on until three evaluation were performed.

Results

No side effects were noted following biopsy and grafting. It was possible to obtain millions of cells in a few weeks and they exhibited similar characteristics to human ones, including: (i) morphology; (ii) strong expression of Nestin and S100A4 proteins; (iii) similar expression of surface markers CD34, CD44 and CD73; (iv) ability to form spheres; (v) cloning and short doubling time; and (vi) expression of proteins specific to neural lineage.

Tests carried out with TGTA revealed deficits in learning and executive functions compared to young dogs from literature (Araujo *et al.*, 2008; Snigdha *et al.*, 2012). One month after injection, grafted dogs displayed stable scores in TGTA tests, while sham dogs showed decreasing performances, as in discrimination test ($p < 0.05$) (Fig. 1A). Differences were not significant for other tests. Regarding ARCAD/EVEC, we observed better scores in grafted animals compared to the baseline, especially for the cognitive competence ($p < 0.05$) (Fig. 1B). Detailed analyses revealed mainly decreased 'Self-control' and 'Learned specific behaviours' parameters.

Discussion

These results suggest that OE-MSCs may be a promising candidate to cope with age-related disorders in dogs. As we observed in previous clinical cases, injection of stem cells into the fourth ventricle seems to quickly reduce emotional and cognitive issues associated with pathological ageing. When cells were grafted in senior dogs, we observed improvement of some clinical issues (e.g. decrease of stereotyped movements, improvement in exploration and capacity to focus attention). Here however, stem cells injection rather seems to reduce cognitive decline.

We only observed minor effects in a limited number of patients, but it should be stated that no side effects have been observed so far. This therapy could be promising but needs to be confirmed in wide clinical studies. Follow-up will continue over 4 months to evaluate if benefits remain stable over time and to ensure that no long-term side effects appear.

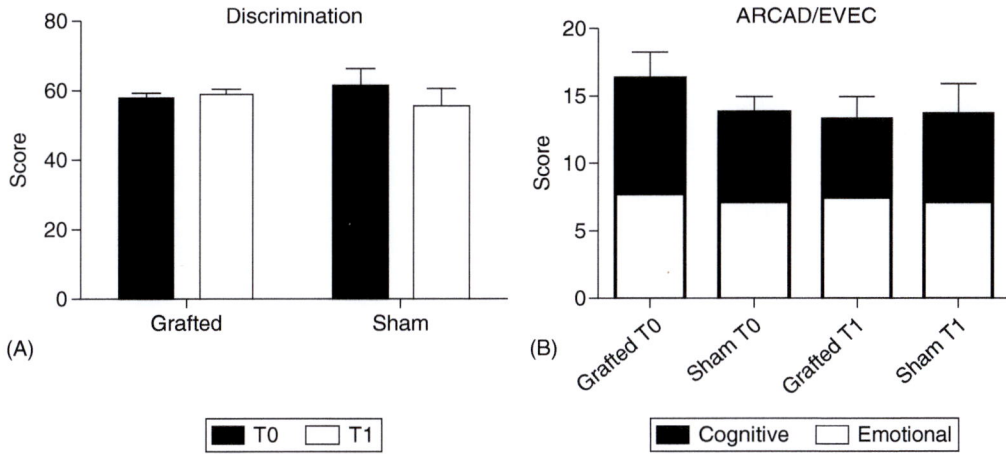

Fig. 1. Assessment of cognitive and emotional abilities before and after graft. Cognitive abilities of dogs were assessed with four TGTA based test. (A) Week score for 'Discrimination' test remained stable for grafted animals, while sham performances decreased. (B) ARCAD/EVEC evaluation revealed stability of emotional competence for both groups, but a decrease of cognitive competence after grafting stem cells.

Conclusion

These results are encouraging for future therapies for pathological ageing in senior dogs. Moreover, collecting of olfactory mucosa tissue is a safe and fast procedure, and tissue could be kept hours before being processed. Cell cultures can multiply in a few weeks, and then could be easily transported before grafting or cryopreservation. These data and the features of these cells support further research towards the development of new cell-based therapy to reduce age-related disorders in dogs.

References

Araujo, J.A., Landsberg, G.M., Milgram, N.W. and Miolo, A. (2008) Improvement of short-term memory performance in aged beagles by a nutraceutical supplement containing phosphatidylserine, Ginkgo biloba, vitamin E, and pyridoxine. *Canadian Veterinary Journal* 49, 379–385.

Colle, M.A., Hauw, J.J., Crespeau, F., Uchihara, T., Akiyama, H., Checler, F., Pageat, P. and Duykaerts, C. (2000) Vascular and parenchymal Abeta deposition in the aging dog: correlation with behavior. *Neurobiology of Aging* 21, 695–704.

Delorme, B., Nivet, E., Gaillard, J., Haupl, T., Ringe, J., Deveze, A., Magnan, J., Sohier, J., Khrestchatisky, M., Roman, F.S., Charbord, P., Sensebe, L., Layrolle, P. and Feron, F. (2010) The human nose harbors a niche of olfactory ectomesenchymal stem cells displaying neurogenic and osteogenic properties. *Stem Cells and Developmemt* 19, 853–866.

Di Trapani, M., Bassi, G., Ricciardi, M., Fontana, E., Bifari, F., Pacelli, L., Giacomello, L., Pozzobon, M., Feron, F., De Coppi, P., Anversa, P., Fumagalli, G., Decimo, I., Menard, C., Tarte, K. and Krampera, M. (2013) Comparative study of immune regulatory properties of stem cells derived from different tissues. *Stem Cells and Development* 22, 2990–3002.

Landsberg, G.M., Nichol, J. and Araujo, J.A. (2012) Cognitive dysfunction syndrome: a disease of canine and feline brain aging. *Veterinary Clinics of North America: Small Animal Practice* 42, 749–768.

Maltman, D.J., Hardy, S.A. and Przyborski, S.A. (2011) Role of mesenchymal stem cells in neurogenesis and nervous system repair. *Neurochemistry International* 59, 347–356.

Mengoli, M., Veron, A.D., Royer, D., Bienboire-Frosini, C., Cozzi, A., Asproni, P. and Pageat, P. (2016) Olfactory stem cells treating age-related behavioural disorders in dogs – a clinical case. ECAWBM. Cascais (Portugal).

Murrell, W., Feron, F., Wetzig, A., Cameron, N., Splatt, K., Bellette, B., Bianco, J., Perry, C., Lee, G. and Mackay-Sim, A. (2005) Multipotent stem cells from adult olfactory mucosa. *Developmental Dynamics* 233, 496–515.

Murrell, W., Wetzig, A., Donnellan, M., Feron, F., Burne, T., Meedeniya, A., Kesby, J., Bianco, J., Perry, C., Silburn, P. and Mackay-Sim, A. (2008) Olfactory mucosa is a potential source for autologous stem cell therapy for Parkinson's disease. *Stem Cells* 26, 2183–2192.

Nivet, E., Vignes, M., Girard, S.D., Pierrisnard, C., Baril, N., Deveze, A., Magnan, J., Lante, F., Khrestchatisky, M., Feron, F. and Roman, F.S. (2011) Engraftment of human nasal olfactory stem cells restores neuroplasticity in mice with hippocampal lesions. *Journal of Clinical Investigation* 121, 2808–2820.

Ould-Yahoui, A., Sbai, O., Baranger, K., Bernard, A., Gueye, Y., Charrat, E., Clement, B., Gigmes, D., Dive, V., Girard, S.D., Feron, F., Khrestchatisky, M. and Rivera, S. (2013) Role of matrix metalloproteinases in migration and neurotrophic properties of nasal olfactory stem and ensheathing cells. *Cell Transplant* 22, 993–1010.

Pageat, P., Veron, A.D., Royer, D., Bienboire-Frosini, C., Asproni, P., Mengoli, M. and Cozzi, A. (2014) Engrafment of seniles dogs with olfactory stem cells: preliminary results for a promising treatment. *Proceedings of Veterinary Behavior Symposium*, Denver, CO, pp. 31–32.

Shafiee, A., Kabiri, M., Ahmadbeigi, N., Yazdani, S.O., Mojtahed, M., Amanpour, S. and Soleimani, M. (2011) Nasal septum-derived multipotent progenitors: a potent source for stem cell-based regenerative medicine. *Stem Cells and Development* 20, 2077–2091.

Snigdha, S., Christie, L.A., De Rivera, C., Araujo, J.A., Milgram, N.W. and Cotman, C.W. (2012) Age and distraction are determinants of performance on a novel visual search task in aged Beagle dogs. *Age (Dordr)* 34, 67–73.

Stamegna, J.C., Girard, S.D., Veron, A., Sicard, G., Khrestchatisky, M., Feron, F. and Roman, F.S. (2014) A unique method for the isolation of nasal olfactory stem cells in living rats. *Stem Cell Research* 12, 673–679.

Tapp, P.D., Siwak, C.T., Estrada, J., Head, E., Muggenburg, B.A., Cotman, C.W. and Milgram, N.W. (2003) Size and reversal learning in the beagle dog as a measure of executive function and inhibitory control in aging. *Learning & Memory* 10, 64–73.

Uccelli, A., Benvenuto, F., Laroni, A. and Giunti, D. (2011) Neuroprotective features of mesenchymal stem cells. *Best Practice & Research Clinical Haematology* 24, 59–64.

Veron, A.D., Mengoli, M., Bienboire-Frosini, C., Royer, D., Asproni, P., Cozzi, A., Girard, S.D., Krestchatisky, M., Roman, F.S. and Pageat, P. (2014) Are nasal stem cells a promising approach in geriatric veterinary medicine? *Proceedings of the IRSEA International* Congress, Apt, France.

Neophilia in Successfully Ageing Dogs: Preliminary Results

Dóra Szabó[1]*, Zsófia Bognár[1], Bianka Stiegmann[1], Renáta Böröczki[1], Lisa Wallis[1], Ádám Miklósi[1,2] and Enikő Kubinyi[1]

[1]Department of Ethology, ELTE, Budapest, Hungary; [2]MTA-ELTE Comparative Ethology Research Group, Budapest, Hungary

Funding: This research was funded by a grant from the European Research Council (ERC) under the European Union's Horizon 2020 research and innovation programme (Grant Agreement No. 680040) and was supported by the János Bolyai Research Scholarship of the Hungarian Academy of Sciences for EK.

Conflict of interest: The authors declare no conflict of interest.

Keywords: behaviour, dogs, cognitive ageing, neophilia, novel object

Introduction

Engaging with novel objects and higher cognitive performance in old age appears connected in humans (Daffner *et al.*, 2007). Based on previous results (Kaulfuß and Mills, 2008) dogs are attracted towards novel objects over familiar ones. Using a simple setup, we investigated whether this phenomenon is also observable in senior dogs.

Material and Methods

We tested dogs over 8 years of age (n=44), free from overt medical conditions and sensory impairment. During a 1-minute long presentation phase, dogs were encouraged to investigate two identical objects. Afterwards, the dogs left the room for 5 minutes, during which one of the objects was replaced with a novel object. Upon returning for the test phase, dogs were released without any specific command, and the dogs' first contact with the objects was recorded.

---

* Corresponding author: szaboodoora@gmail.com

Results

Twenty-eight dogs manipulated at least one of the objects during this phase (the remaining dogs were passive or oriented towards the humans). Dogs contacted the novel object more frequently than the familiar object (Wilcoxon signed-rank test: $z=-2.197$, $p<0.05$).

Conclusion

Neophilia was evident in normally ageing dogs, even after a very brief presentation and a short break. Our simple protocol could be included in the evaluation of cognitive state of dogs in the veterinary practice; multiple trials can be carried out within minutes, and without pre-training. Future studies should investigate whether the loss of ability to detect novel objects may be an indication of pathological cognitive decline in senior dogs, as it is in humans.

References

Daffner, K.R., Chong, H., Riis, J., Rentz, D.M., Wolk, D.A., Budson, A.E. and Holcomb, P.J. (2007) Cognitive status impacts age-related changes in attention to novel and target events in normal adults. *Neuropsychology* 21(3), 291–300.

Kaulfuß, P. and Mills, D.S. (2008) Neophilia in domestic dogs (*Canis familiaris*) and its implication for studies of dog cognition. *Animal Cognition* 11(3), 553–556.

The Attachment Bond to People in Domestic Dogs: Does It Exist Already in Puppies?

Chiara Mariti[1]*, Lorenzo Lenzini[1], Beatrice Carlone[1], Marcella Zilocchi[1], Beatrice Caverni[2] and Angelo Gazzano[1]

[1]Dep. Veterinary Sciences, University of Pisa, Pisa, Italy; [2]Associazione Indaco, Pisa, Italy

Conflict of interest: The authors declare no conflict of interest

Keywords: Ainsworth Strange Situation Test, attachment, behaviour, bond, dog, puppy

Introduction

Adult dogs can develop relationships with their owners that fulfil the requirements of an attachment bond: proximity seeking, protest at an involuntary separation, and secure base effect (Mariti *et al.*, 2013). The dog–owner relationship appears to have special characteristics allowing dogs to create an attachment. A previous study found that the relationship between adult dogs living in the same household does not fulfil the same requirements. Dogs tested using the Ainsworth Strange Situation Test (ASST) (Ainsworth and Bell, 1970) preferred a strange person over the cohabitant dog (Mariti *et al.*, 2014). When adult dogs are tested with their mother versus another older female dog, they show preference towards their mother. This result suggests that the mother–puppy bond can be maintained, provided they live in the same environment (Mariti *et al.*, 2017). Also, puppies can form an appropriate attachment with their mothers (Prato-Previde *et al.*, 2009). Our hypothesis was that puppies could develop an attachment to the first person they encounter (e.g. the owner of the bitch). The aim of this study was to investigate whether attachment to people already exists in puppies.

---

* Corresponding author: chiara.mariti@unipi.it

Materials and Methods

Fourteen puppies, seven males and seven females, 58–60 days old, from five litters of different breeds (Australian Kelpies, Bearded Collies, Border Collies, Flat-coated Retrievers and Labrador Retrievers) enrolled into the study. All puppies were born and lived in a home environment. The criterion of living at home was selected to ensure that all puppies have the opportunity to spend time and interact with the presumed attachment figure. The presumed attachment figures (breeders or owners) and the strangers were all women of similar age (25–35 years old).

Puppies were tested using the ASST, a behavioural test in which the subject, usually a child or an animal, is tested with a stranger and a presumed attachment figure (e.g. the mother or the owner). The ASST is carried out in a room that is unknown to the tested subject, equipped with toys and chairs for participants (Fig. 1). The test includes several steps: (i) habituation to the test environment; (ii) attachment figure and the subject spending time together in the same environment but not interacting; (iii) the stranger enters the area and interacts with the caregiver who then leaves the room; (iv) the stranger and the puppy spend time in the environment alone; (v) the caregiver enters the room again and interacts with the subject; (vi) both the caregiver and the stranger leave the room, the subject stays alone; (vii) the stranger comes back into the room; and (viii) the caregiver returns to the room and interacts with the subject while the stranger leaves. In this study, behaviours of puppies in steps vi, vii and viii were recorded. Results were analysed using a Wilcoxon test ($p<0.05$ for social behaviours, $p<0.0167$ for non-social behaviours using Bonferroni correction for the three comparisons).

Results

When left alone, puppies showed distress behaviours such as yelping, increased activity, and staying close to the door. These behaviours reduced in the presence

Fig. 1. The dispersal of toys and chairs (one for the breeder and one for the stranger) in the experimental room where the Ainsworth Strange Situation Tests were performed. A puppy is displaying proximity to the owner and individual play.

of the attachment figure or the strangers. The puppies did not show any preference towards the attachment figure or the stranger.

Discussion and Conclusions

The findings of this preliminary study suggest that dog puppies do not develop an attachment with their caregivers by the age of 2 months, as demonstrated by the lack of preference towards the caregiver or the stranger. Based on results of attachment studies in adult dogs towards their caregivers (Mariti *et al.*, 2013) it appears that a longer time spent together is needed for this attachment to develop. Also, it is possible that the primary attachment between the mother and the puppies hinders the establishment of other bonds at the age of 2 months. The distress behaviours that puppies showed during the separation and isolation phase of the study were similar to those reported in the literature of dogs and other species.

Further studies should focus on the relationship between puppies and adopting owners, to assess the time needed for establishing a strong and healthy attachment, and which factors can facilitate its formation.

References

Ainsworth, M.D. and Bell, S.M. (1970) Attachment, exploration and separation – illustrated by behaviour of one year olds in a strange situation. *Child Development* 41, 49–67.

Mariti, C., Ricci, E., Zilocchi, M. and Gazzano, A. (2013) Owners as a secure base for their dogs. *Behaviour* 150, 1275–1294.

Mariti, C., Carlone, B., Ricci, E., Sighieri, C. and Gazzano, A. (2014) Intraspecific attachment in adult domestic dogs (*Canis familiaris*): preliminary results. *Applied Animal Behavior Science* 152, 64–72.

Mariti, C., Carlone, B., Votte, E., Ricci, E., Sighieri, C. and Gazzano, A. (2017) Intraspecific relationships in adult domestic dogs (*Canis familiaris*) living in the same household: a comparison of the relationship with the mother and an unrelated older female dog. *Applied Animal Behavior Science*. Available at: https://doi.org/10.1016/j.applanim.2017.05.014.

Prato-Previde, E., Ghirardelli, G., Marshall-Pescini, S. and Valsecchi, P. (2009) Intraspecific attachment in domestic puppies (*Canis familiaris*). *Journal of Veterinary Behavior* 4, 89–90.

Living with and Loving a Pet with Behaviour Problems: The Impact on Caregivers

KRISTIN BULLER[1] AND KELLY C. BALLANTYNE[2]*

[1]Clinical Social Worker, Chicago, Illinois, USA; [2]University of Illinois College of Veterinary Medicine, Chicago, Illinois, USA

Conflict of interest: The authors report no conflicts of interest.

Keywords: behaviour, companion animals, human–animal bond

Introduction

Many studies have investigated the impacts of behaviour problems on companion animals but few have investigated the impacts of these problems on their caregivers. Studies in human medicine show that caring for mentally ill family members has several impacts on the caregiver's life, and caregivers of mentally ill companion animals may experience similar challenges. The objectives of this study were to provide a detailed and impartial view of the caregiver's experience as well as to inform further research and support.

Materials and Methods

A convenience sample of 63 pet owners took part in a survey. Responses were analysed using thematic analysis, a qualitative method used for identifying, analysing and reporting patterns within data.

Results

Survey respondents reported several impacts on their lives due to their pet's behavioural condition(s). These included impacts on day-to-day life such as the

---

* Corresponding author: kcmorgan@illinois.edu

time required for management and training, difficulty exercising their pet, and limitations on where they could go and who could visit. This impacted household relationships as well as family and friends. Most survey respondents reported feeling strongly bonded with their pet while also reporting a range of negative emotional responses to their pet's behavioural problem.

Conclusion

Caregivers of mentally ill companion animals are impacted in several ways. While animal health professionals can serve as a source of support, additional support resources could include individual or group counselling with a qualified therapist or social worker. Establishing a collaborative relationship with such a professional could ensure that both the pet's and the owner's needs are met when managing problem behaviours.

The Personality of Dogs and Cats Living in the Same Household: A Multivariate Model Study

LAURA MENCHETTI, SILVIA CALIPARI AND SILVANA DIVERIO*

Laboratory of Ethology and Animal Welfare (LEBA) Department of Veterinary Medicine, Perugia University, Italy

Conflict of interest: The authors declare no conflict of interest.

Keywords: personality traits, Five-Factor model, multi-pet households, multivariate analysis

Introduction

Personality is the result of interactions between genes, environmental and experiential factors. Dogs and cats living in the same household are a valuable experimental model to evaluate the effect of factors affecting personality traits.

Material and Methods

A total of 1270 owners of dogs and cats in the same household answered to a multiple-choice questionnaire, collecting data on their pets' personality, demographic features and management habits. Principal component analysis and multivariable regression models were used.

Results

Five personality traits link dogs and cats: Sociability, Calm, Protectiveness, Neuroticism and Fear. Dogs scored higher in Sociability and Protectiveness whilst scoring lower in Neuroticism than cats (p<0.001). Age showed the same effects in Sociability and Calm of dogs and cats (p<0.01). Gender affected Sociability

* Corresponding author: silvana.diverio@unipg.it

and Neuroticism in cats ($p<0.001$) and Fear in dogs ($p<0.05$). Neutering status, holding constant the other factors, only modulated dogs' Fear ($p<0.05$). Age of owners affected Protectiveness and Calm of dogs ($p<0.05$) and Neuroticism of cats ($p<0.01$). Age of acquisition affected Sociability, Protectiveness and Fear of dogs ($p<0.05$) and no trait of cats. Management habits modulated all the traits of dogs' personality ($p<0.05$), while the presence of other animals in the house influenced Sociability, Neuroticism and Fear of the cat ($p<0.001$).

Conclusion

Although cats and dogs share the same social and physical environment, their personality showed a prevailing cross-species plasticity. The multivariable models highlight confounding effects and interactions between factors suggesting potential issues of coexistence between cats and dogs. Moreover, multivariable models can have predictive significance providing another use in clinical and practical settings of veterinary medicine.

A Comparison of Attitudes and Coping Strategies of Small Animal Veterinarians Towards Behaviour Problems of Companion Dogs in Israel and Brazil

Yon Alexandre Raileanu[1], Liat Baht-Segal[2], Joseph Terkel[3] and Noa Harell[2]*

[1]Faculdades Metropolitanas Unidas University, Sao Paolo, Brazil; [2]Hebrew University, Jerusalem, Israel; [3]Tel-Aviv University, Tel-Aviv, Israel

Conflict of interest: The authors declare no conflict of interest.

Keywords: veterinary behaviour medicine, behaviour problems, companion dogs

Introduction

Over the past three decades, the responsibility for diagnosing and treating behavioural disorders of companion animals has been shifting towards the veterinary profession (Mills, 2003; Juarbe-Diaz, 2008; Harell, 2012). To date, many veterinarians still avoid diagnosing and treating many behaviour problems of their patients, and may not apply strict criteria when deciding to refer these patients on to other professionals (Hetts *et al.*, 2004; Landsberg *et al.*, 2008).

In an attempt to evaluate the current position of the veterinary profession regarding its approach to behavioural medicine, two surveys were carried out in Israel (2012–2015) and Brazil (2015–2017). The aim was to obtain data on the following subjects: frequency of behavioural problems encountered by veterinary surgeons in first opinion practices; veterinarians' perception of their responsibility for addressing these complaints; choices made by veterinarians in treating vs referring cases to other veterinary and non-veterinary professionals; and the level of education in behavioural medicine veterinarians receive.

* Corresponding author: vetbehaviorclinic@gmail.com

©S. Denenberg 2017. *Proceedings of the 11th International Veterinary Behaviour Meeting*
(ed. S. Denenberg)

Our working hypotheses were: that most veterinarians do not diagnose and treat behaviour problems in their practices, but refer them to non-veterinary professionals, veterinary behaviourists or other resources; that in general, veterinarians do offer behavioural advice for puppies and kittens and that this advice is mainly given during consultations scheduled for other purposes (Patronek and Dodman, 1999); that the reasons for choosing to refer rather than treat behaviour problems in-house include lack of theoretical knowledge and practical experience and lack of time (Horwitz, 2008); and the perception that most behaviour problems do not lie within the responsibility of the veterinary practitioner – a perception that has been shown to be prevalent in previous studies carried out in other countries (Patronek and Dodman, 1999; Fatjo *et al.*, 2006; Soares *et al.*, 2010).

Materials and Methods

Two hundred and twenty-one veterinary surgeons in Israel (n=91) and Brazil (n=130) responded to online questionnaires. Participants were male and female veterinarians of ages 25 and up, from both urban and rural areas. The questionnaires were sent out via email to small animal practitioners (Israel) and via email to veterinary practitioners, clinical directors of veterinary clinics and veterinary course coordinators of different universities to forward for the practitioners and professors (Brazil). The items in the questionnaires included demographic information; information on formal education in behavioural medicine; behavioural complaints and problems encountered during routine clinical work; and information on different possible approaches to these complaints. Results were analysed using Pearson Chi-square test and Fisher's Exact test (both Israel and Brazil) with $p < 0.05$.

Results

Most participants (97% in Israel, 86% in Brazil) are regularly consulted on behavioural concerns in dogs. Seventy per cent of respondents in Israel perceive behaviour problems of their patients to be part of their responsibility. Forty-two per cent of respondents in Israel and 99% in Brazil had no formal training in behavioural medicine as veterinary students. However, 67% of the Israeli respondents and 32.2% in Brazil still chose to treat at least part of the behavioural problems of their patients. The behaviour problems most commonly addressed by veterinarians in Israel were inappropriate elimination (55%), noise sensitivity (44%) and repetitive behaviours (43%). In Brazil, the most common behavioural problems addressed by replying participants were excessive self-licking (69.2%), separation anxiety (52.3%), aggression (49.2%) and excessive vocalisation (42.3%).

Furthermore, participants in Israel reported that they advise clients on puppy behaviour (91%) and senior dog behaviour complaints (67%). Of those veterinarians who do not treat behaviour problems in-house, only 5.5% replied that they do not view treating behaviour problems as part of routine clinic work. Others responded that they lack knowledge and experience (48.4% in Israel, 10% in Brazil), did not receive education in behavioural medicine (16.5%, Israel) or were lacking in time to treat these problems (7.7%, Israel). In Israel, 97.8% of participants believed that it is important to receive training in behavioural medicine in veterinary school. In Brazil, 21.5% of participants report that although they lack knowledge and experience, the behavioural cases they are consulted on motivated them to search for appropriate treatments for these problems.

Discussion

The results of these surveys show that in both countries, veterinary surgeons tend to treat some types of behavioural problems, whereas other types of problems are referred to non-veterinary professionals. Differences also exist between Israel and Brazil regarding the types of behaviour problems or complaints addressed more commonly by veterinarians. Here, we offer some possible explanations for these differences. It is possible that veterinarians feel more confidence in treating problems that are thought to be more responsive to medication (psychoactive or other), rather than behavioural and environmental modification. These include noise sensitivity, repetitive behaviours and separation-related problems. Also, veterinarians may feel more competent to treat cases in which another body system is involved in the behavioural presentation, such as inappropriate elimination. Culture may also play a part in other differences in approach to various problems between the two countries.

In both countries, veterinarians feel they lack knowledge and experience in behavioural medicine. Another significant finding is the correlation between receiving formal education and the confidence to treat behavioural cases, as shown in the striking differences between the two countries in the comparative study.

Conclusions

The correlation between receiving formal education and the confidence to treat behavioural cases in-clinic emphasises the need to offer and improve the level of education in behavioural medicine for both students and veterinary surgeons. The second implication of this study is that veterinarians may still view a large part of behavioural problems as non-veterinary, thereby referring these to other professionals. Since certain presentations (e.g. aggression and house-soiling) require veterinary attention, students in veterinary schools should be appropriately educated as part of the veterinary curriculum. In turn, this education will extend the scope of behavioural problems treated by veterinary surgeons.

References

Fatjo, J., Ruiz De La Torre, J.A. and Manteca, X. (2006) The epidemiology of behavioural problems in dogs and cats: a survey of veterinary practitioners. *Animal Welfare* 15, 179–185.
Harell, N. (2012) A new perspective on teaching veterinary behavioral medicine. In *Proceedings of Veterinary Behavior Symposium*, San Diego, California, p. 58.

Hetts, S., Estep, D. and Heinke, M.L. (2004) Behavior wellness concepts for general veterinary practice, *Journal of the American Veterinary Medical Association* 225(4), 506–513.

Horwitz, D.F. (2008) Managing pets with behavior problems: realistic expectations. *Veterinary Clinics of North America: Small Animal Practice* 38(5), 1005–1021.

Juarbe-Diaz, S.V. (2008) Behavioral medicine opportunities in North American colleges of veterinary medicine: a status report. *Journal of Veterinary Behavior; Clinical Applications and Research* 3, 4–11.

Landsberg, G.M., Shaw, J. and Donaldson, J. (2008) Handling behavior problems in the practice setting. *Veterinary Clinics of North America: Small Animal Practice* 38, 951–969.

Mills, D.S. (2003) Medical paradigms for the study of problem behaviour: a critical review. *Applied Animal Behavior Science* 81(3), 265–277.

Patronek, G.J. and Dodman, N.H. (1999) Attitudes, procedures, and delivery of behavior services by veterinarians in small animal practice. *Journal of the American Veterinary Medical Association* 215(11), 1606–1611.

Soares, G.M., Mattos de Souza-Dantas, L., D'Almeida, J.M. and Paixao, R.L. (2010) Epidemiologia de problemas comportamentais em cães no Brasil: inquérito entre médicos veterinários de pequenos animais. *Ciência Rural* 40(4), 873–879.

The Latin Owner: Profiles, Perceptions and Attitudes of Italian Cat and Dog Owners Towards their Pet

Federica Pirrone[1]*, Ludovica Pierantoni[2] and Mariangela Albertini[1]

[1]Department of Veterinary Medicine, University of Milan, Milan, Italy;
[2]CAN (Comportamento Animale Napoli) s.s.d.r.l., Naples, Italy

Conflict of interest: The authors declare no conflict of interest.

Keywords: pets, ownership, perception

Introduction

Nowadays animal companionship is an integral aspect of life in Europe, with approximately 81 million registered dogs and 99.2 million cats[1]. This research aims to identify characteristics of dog and cat-owning households from a large cross-sectional internet-based survey in Italy.

Materials and Methods

Owners over 18 years old were asked information about themselves, their dogs, cats and their relationship with their pets. Data were analysed using Pearson's χ^2 tests and logistic regressions (SPSS).

Results

A total of 3298 owners completed the survey, 31.8% and 40.3% of whom owned dogs and cats respectively, and 72.8% both. People aged 18 to 30 years

---

* Corresponding author: federica.pirrone@unimi.it

©S. Denenberg 2017. *Proceedings of the 11ᵗʰ International Veterinary Behaviour Meeting* (ed. S. Denenberg)

were more likely to own a dog than older respondents. Compared to cat owners, dog owners were more likely to believe that their pets considered them to be conspecific group members, rather than 'only humans'. Dogs were more likely to be purebreds adopted for companionship. Cats were more likely to be mixed breeds adopted because they needed a home. Dog owners were significantly more likely to rate other owners as an important source of information regarding handling and training than cat owners. Despite a similarly high prevalence of reported intraspecific aggression and noise reactivity among dogs and cats, dog ownership significantly increased the likelihood of the owner's actual willingness to change a pet's behaviour. Cat ownership increased the likelihood of considering castration as an option to correct behaviour.

Conclusion

These results may be useful in helping behaviour consultants to address population changes in terms of human–pet bonds, and plan preventive and treatment strategies.

Note

[1] Statista 2015. Available at https://www.statista.com/statistics/198100/dogs-in-the-united-states-since-2000/

Coping Strategies in Dogs with Impaired Social Functioning Towards Humans

Tiny De Keuster[1,2]*, Joke Monteny[3] and Katrien Verschueren[4]

[1]Faculty of Veterinary Medicine, Ghent University, Merelbeke, Belgium; [2]Behavioral Referrals, Lovendegem, Belgium; [3]Hondinform, Heuvelland, Belgium; [4]Living Statistics, Gent, Belgium

Conflict of interest: The authors declare no conflict of interest.

Keywords: human–dog, benign interactions, coping strategy, prevention, aversive education

Introduction

This study describes behavioural strategies in dogs with impaired social functioning within the context of benign human interaction, and investigating if these behavioural strategies can be useful in assessing a dog's social competence in human–dog interactions.

Materials and Methods

Data from 200 dogs' cases was collected from authors' referral centre database, and based on questionnaire, video footage and behavioural examination. Dogs' coping strategies were categorised into 'Subtle', 'Defuse' and 'Escalation'. Based on observations, additional categories were created including 'Human-directed', 'Object-directed' and 'Motor activity'. For each individual dog, coping strategies were scored as 0 (absence) or 1 (presence). A Global Score was calculated for each group.

* Corresponding author: tiny.dekeuster@ugent.be

Results

Positive scores were found in the 'Defuse' group for 179 dogs (89.5%). In the 'Escalation' group results showed biting in 86 (43%), snapping in 114 (57%) and growling in 131 dogs (65.5%). Further, positive scores were found in the 'Human-directed' group for 177 dogs (88.5%), 134 dogs (67%) in the 'Object-directed' group and 158 dogs (79%) in the 'Motor activity' group. A negative correlation (−0.36; $p<0.0001$) was found between Global Scores in 'Human-directed' and 'Defuse' strategies, indicating that dogs displaying more human-directed strategies display less conflict-defusing strategies. Application of aversive punishment was more frequent in owners of dogs with high Global Human scores. This resulted in significant increases in 'Escalation', including growling (86%), snapping (85%) or biting (64%).

Conclusion

Socially impaired dogs may respond with strategies not commonly described as an indication of social conflict. This lack of awareness may increase the use of punishment leading to escalation.

Predictors of Gaze-directed Attention in Dogs

Zsófia Bognár, Ivaylo Borislavov Iotchev and Enikő Kubinyi*

Department of Ethology, Eötvös Loránd University, Budapest, Hungary

Funding: This study was funded by a grant from the European Research Council (ERC) under the European Union's Horizon 2020 research and innovation program (Grant Agreement No. 680040) and from the Bolyai Foundation of the Hungarian Academy of Sciences.

Conflict of interest: The authors declare no conflict of interest.

Keywords: behaviour, dogs, attention, gazing

Introduction

A growing body of evidence suggests that eye-contact and gaze-following are important parts of the dog's social repertoire, but little is known about the factors facilitating attention towards the eyes and face. The aim of this study was to investigate factors that influence dogs' gaze-directed attention.

Material and Methods

We used a semi-automated image presentation of humans and dog heads, either facing the observer (portrait) or facing away (profile). These stimuli were presented to 38 pet dogs on a screen (1) in a spontaneous looking condition without food and (2) in the presence of a food reward in front of the screen. In the latter condition, we assumed that dogs might be more hesitant to approach the food if a dog portrait is facing them depending on breeds.

---

* Corresponding author: eniko.kubinyi@ttk.elte.hu

Results

Dogs looked longer at portraits than at profiles, and looked longer at images of other dogs than of humans. Female dogs and dogs of brachycephalic head-shape looked longer at the images in both conditions, and approached the food reward slower. Looking time was longest for dogs belonging to non-cooperative breeds in the without food condition and dogs belonging to cooperative breeds in the with food condition. Finally, old dogs looked longer and approached food slower.

Conclusion

Our results suggest that in the dog gaze-directed attention is driven by mechanisms similar to those found in humans and by factors previously found to aid the dog's understanding of human gestures. We also encourage further exploration of the relationship of this social communicative behaviour with gender, development and ageing.

The Effect of Paw Preference on Problem-solving Ability in Cats: Preliminary Results

Sevim Isparta[1]*, Yasemin Salgirli Demirbas[2] and Gonçalo Da Graça Pereira[3]

[1]Department of Genetics, Faculty of Veterinary Medicine, Ankara University, Ankara, Turkey; [2]Department of Physiology, Faculty of Veterinary Medicine, Ankara University, Ankara, Turkey; [3]Centro para o Conhecimento Animal, Algés, Portugal

Conflict of interest: The authors declare no conflict of interest.

Keywords: cat, paw preference, problem solving, laterality

Introduction

Many authors have suggested that behavioural lateralisation increases neural capacity to carry out simultaneous processing (Vallortigara and Rogers, 2005). The aim of this study is to investigate the relationship between strength of paw preference and problem-solving ability in domestic cats.

Materials and Methods

Fourteen cats were tested in a kennel environment, at a cattery. Informed consents were obtained from the cat owners prior the study. A cat toy on a wand was presented to each cat to determine the paw preference. A problem-solving test, which included three different steps (T1, T2, T3), was administered to each cat. Canned cat food was used as a reward for test steps.

---

* Corresponding author: sevimisparta93@gmail.com

Results

A significant correlation was found between strength of paw preference in play and T2 (r=0.842; p<0.01). Greater incidence of the ambilateral preference was observed in T1 (50%) in comparison to play (33.3%), T2 (36.4%) and T3 (0%). Success rates significantly decreased among test sessions (Friedman test, p<0.01). Stronger paw preference was observed in T3 in comparison to T1. However, this difference was not found to be statistically significant (Wilcoxon test, p=0.083).

Discussion

Since T1 was the first step of the problem-solving test, it is possible that it may cause more emotional reactivity, and thus, more ambilateral preferences in comparison to other steps. The preliminary findings were consistent with the previous study suggesting that stronger lateralisation was apparent on more complex manipulative tasks in cats (Wells and Millsopp, 2009).

References

Vallortigara, G. and Rogers L.J. (2005) Survival with an asymmetrical brain: advantages and disadvantages of cerebral lateralization. *Behavioural and Brain Sciences* 28, 575–589.

Wells, D.L. and Millsopp, S. (2009) Lateralized behaviour in the domestic cat, *Felis silvestris catus*. *Animal Behaviour* 78(2), 537–541.

The Impact of Transportation-related Anxiety on the Quality of Pre-anaesthesia in Cats: A Preliminary Study

Juan Argüelles[1]*, Mónica Echaniz[1], Jon Bowen[2], Paula Calvo[3] and Jaume Fatjó[3]

[1]Centro Veterinario Integral La Cañada, Valencia, Spain; [2]Queen Mother Hospital for Small Animals, The Royal Veterinary College, Hatfield, UK; [3]Departament of Psychiatry, School of Medicine, Autonomous University of Barcelona, Barcelona, Spain

Conflict of interest: The authors declare no conflict of interest.

Keywords: anxiety, cats, transportation, anaesthesia

Introduction

In human medicine, there is evidence for a relationship between perioperative anxiety and anaesthetic requirements. Similarly, veterinary surgeons often indicate certain difficulties to induce anaesthesia in agitated and fearful patients. Transportation is a stressful situation for most cats. The aim of this study was to explore the relationship between transportation-related anxiety and anaesthetic requirements.

Materials and Methods

Fifty-four cats (27 females and 27 males) undergoing an elective surgery in a single veterinary practice were included. For ethical reasons, before the study the caregivers of all patients were given a protocol to reduce anxiety during transport, including handling instructions and the application of F3 synthetic pheromone.

* Corresponding author: juanin73@gmail.com

©S. Denenberg 2017. *Proceedings of the 11th International Veterinary Behaviour Meeting* (ed. S. Denenberg)

We included in group 1 those patients whose caregivers did not comply with the protocol, and in group 2 those patients whose caregivers complied. Parameters to assess anaesthesia included time to reach sedation, urinary and blood cortisol, heart rate, respiratory rate and induction agent (Propofol) dose. A Mann–Whitney test was used to analyse the results.

Results

Forty-nine cases completed the study: 27 in the treatment group and 22 in the control group. A significantly shorter time to reach sedation was found in the treatment group compared to the control group. A trend for a lower dosage of Propofol in the treatment group was also observed.

Conclusions

Reducing perioperative anxiety using behaviour modification and pheromone appear to have a positive effect on the quality of anaesthesia. Additionally, lower anxiety can lead to increased safety during anaesthesia as a result of faster sedation and lower doses of anaesthetic agents.

Playful Activities Post-learning Improve Retraining Performance a Year Later in Labrador Retriever Dogs (*Canis lupus familiaris*)

Nadja Affenzeller[1,2]* and Helen Zulch[1]

[1]*Animal Behaviour, Cognition and Welfare Group, School of Life Sciences, University of Lincoln, Lincoln, UK;* [2]*Department of Companion Animals, Clinical Unit of Internal Medicine Small Animals, University of Veterinary Medicine Vienna, Vienna, Austria*

Conflict of interest: The authors declare no conflict of interest.

Keywords: dog, memory, play, rest, training performance

Introduction

Arousing and emotional situations are known to improve cognitive performance and memory of events. It is thought that beta-adrenergic activation and the release of specific stress hormones enhance memory consolidation and lead to an increase in remembering through facilitation of memory recall (McGaugh, 2000).

This has been shown in humans, non-human primates, rodents (Cahill *et al.*, 2001; Roozendaal *et al.*, 2001) and most recently in dogs (Affenzeller *et al.*, 2017). It was demonstrated that training performance can be enhanced in Labrador Retriever dogs engaged in 30 minutes of playful activities (PA) immediately after acquiring a two-choice object discrimination task when compared with allowing dogs to rest (RP) post-training. Techniques which could enhance memory and improve training efficacy for specific tasks would be tremendously valuable, especially in dogs, who are extensively trained to aid humans.

* Corresponding author: naffenzeller@lincoln.ac.uk

Materials and Methods

In a previous study (with 16 dogs) after initial acquisition of the task on day 1, either a playful activity intervention (n=8) or a resting period (n=8) took place, lasting for 30 minutes. All dogs then underwent the same procedure 24 hours later.

In the current study 11 of these Labrador Retrievers ranging from 1 to 11 years of age were re-trained in the 2-choice discrimination paradigm (Fig. 1) more than 1 year after initial task acquisition. Dogs were re-trained based on their previous object and group allocation, until training criterion was met (success rate ≥80% in two consecutive sessions). To avoid a location and experimenter bias all dogs where re-tested in a different room and re-trained by two different experimenters.

In the current study, factors including type of intervention, experimenter, training performance in the previous study (numbers of trials needed to re-learn the task after 24 hours), average heart rate (during intervention and during re-training) and cortisol level data (after intervention) on number of trials and errors to meet re-training criterion were subjected to a multiple factor/covariate General Linear Model analysis.

Results

A ceiling effect was observed; 3 of 5 dogs in the PA group (Fig. 2) and 1 of 6 in the RP group (Fig. 3) reached criterion within the first two sessions of training.

Errors made in Session 1 and 2 were significantly affected by the type of intervention (p=0.021, F=7.73, DF 1, 10). Dogs from the PA group (n= 5) made

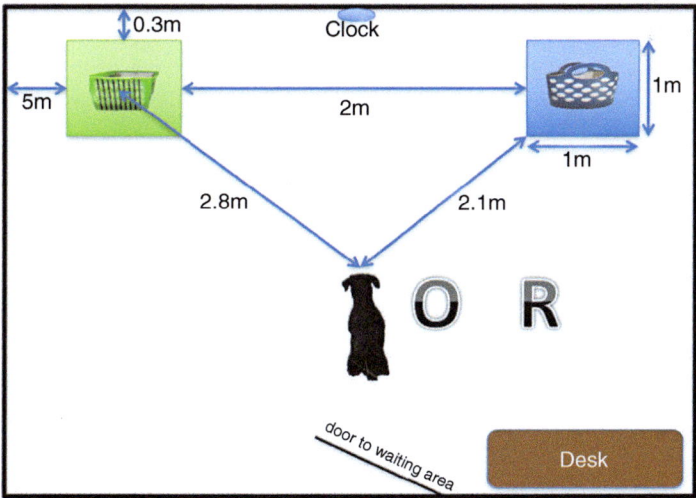

Fig. 1. Setup and dimensions of the testing area. O, designated area of the owner; R, designated area of the researcher.

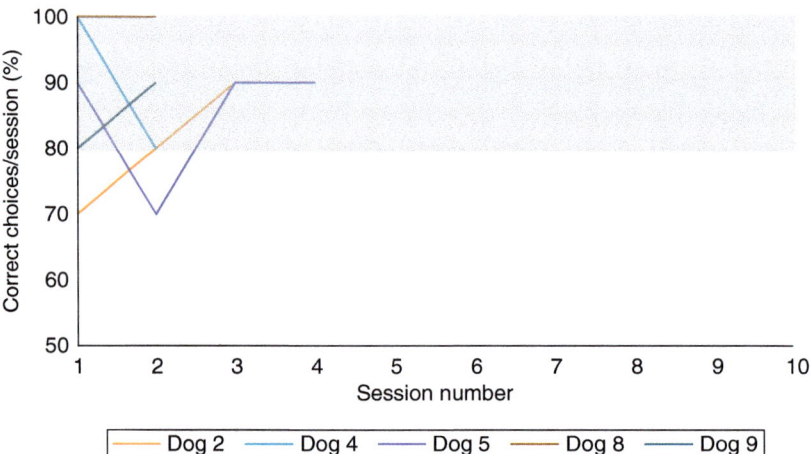

Fig. 2. Individual training curves of dogs assigned to the playful activities group. The grey rectangle highlights the successful training criteria area (≥80% in two consecutive sessions).

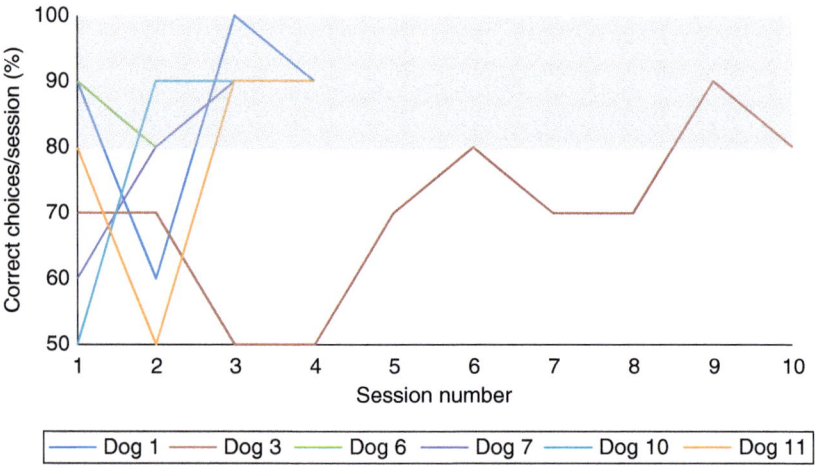

Fig. 3. Individual training curves of dogs assigned to the resting group. The grey rectangle highlights the successful training criteria area (≥80% in two consecutive sessions).

significantly fewer errors in training sessions 1 and 2 (mean number of errors PA group: 3.6, SD 2.1; RP group: 6.5, SD 1.4; 2 sample t-test: $t(6)=-2.6$ $p=0.037$, effect size $d=1.6$) than dogs from the RP group ($n=6$) (Fig. 4).

No significant effect was found of experimenter, previous training performance, average heart rate (during intervention or during re-training), cortisol levels post intervention or type of intervention on absolute numbers of trials and absolute numbers of errors needed to re-learn the task >1 year later (mean number of trials PA group: 22.6, SD 6.9; RP group: 33.2, SD 19).

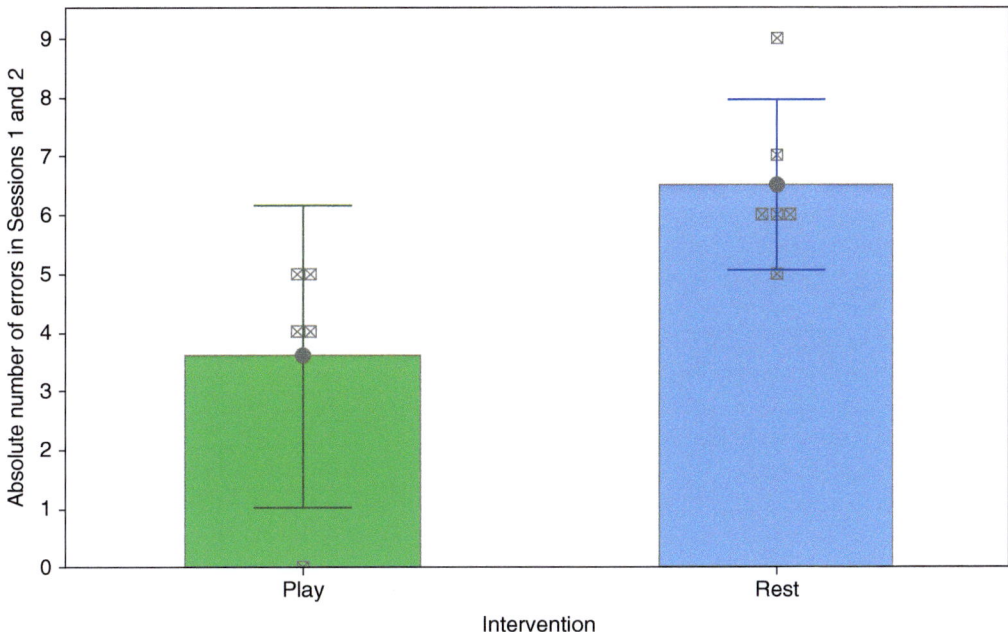

Fig. 4. Individual value plot and bar chart of absolute number of errors made in Sessions 1 and 2 to reach training criterion. Playful activity group (Play) represented in green, resting group (Rest) represented in blue. Interval bars represent 95% confidence interval for the mean. Squares represent individual dogs.

In addition, there was no significant difference between groups with respect to the length of time passed before retesting (retesting interval PA mean 13.2 months, SD 2.7; RP 14.7 months, SD 2.2; 2-sample t-test p>0.1) and mean heart rate during re-training (PA mean 101 bpm, SD 15; RP mean 108 bpm, SD 16; 2-sample t-test p>0.1).

Discussion

To the author's knowledge this is the first evidence that post-training activity may influence memory in dogs more than one year after initial task acquisition. Playful activities for 30 minutes post-learning affect long-term memory not only after 24 hours (Affenzeller *et al.*, 2017) but also over 1 year later.

It needs to be pointed out that the effect of the intervention was only significant for the absolute number of errors made in training session 1 and 2. It is hypothesised that the previously published positive effect of playful activities on absolute number of trials needed to successfully re-learn the task could not be replicated due to an observed ceiling effect. Three of 5 dogs in the PA group versus one of 6 dogs in the RP group reached training criteria (≥80% success rate) within the first 2 trials, obscuring a potentially significant effect of the intervention and other factors on overall training performance 1 year later. In addition to that, only 11 out of the original 16 dogs were available for re-training 1 year later and this decrease in sample size might have affected statistical power.

A limiting factor in the previous study was that the same researcher conducted both the training and the interventions, which might have led to the researcher (or the location) being positively associated with treats and playful interactions. In addition, the researcher was not blinded for the type of intervention, and despite control measures in place potentially inadvertently cued the dog when re-training 24 hours later. For this reason, all dogs were re-trained 1 year later in a different location by two unfamiliar researchers unaware of the intervention group, with no effect seen on training performance in either group.

In addition, there was no difference observed in mean heart rate data during re-training suggesting that beta-adrenergic activation and hence arousal should not have affected training performance in either group (Smeets *et al.*, 2008).

Future studies should incorporate a control group engaged in exercise only without an emotional component after training as it has been shown that acute exercise positively impacts on memory (Roig *et al.*, 2013; Snigdha *et al.*, 2014).

Conclusion

These preliminary results warrant further investigations into how memory recollection and training performance in dogs can be optimised. A better understanding of efficacious interventions in different training tasks (such as odour detection in working dogs) would be of tremendous practical use both in the professional and the private sector of dog training.

References

Affenzeller, N., Palme, R. and Zulch, H. (2017) Playful activity post-learning improves training performance in Labrador Retriever dogs (*Canis lupus familiaris*). *Physiology & Behavior* 168, 62–73.

Cahill, L., McGaugh, J. and Weinberger, N. (2001) Opinion: the neurobiology of learning and memory: some reminders to remember. *Trends in Neurosciences* 24, 578–581.

McGaugh, J.L. (2000) Memory – a century of consolidation. *Science* 287(5451), 248.

Roig, M., Nordbrandt, S., Geertsen, S. and Nielsen, J. (2013) The effects of cardiovascular exercise on human memory: a review with meta-analysis. *Neuroscience & Biobehavioral Reviews* 37, 1645–1666.

Roozendaal, B., Phillips, R., Power, A., Brooke, S., Sapolsky, R. and McGaugh, J. (2001) Memory retrieval impairment induced by hippocampal CA3 lesions is blocked by adrenocortical suppression. *Nature Neuroscience* 4 (12), 1169–1171.

Smeets, T., Otgaar, H., Candel, I. and Wolf, O. (2008) True or false? Memory is differentially affected by stress-induced cortisol elevations and sympathetic activity at consolidation and retrieval. *Psychoneuroendocrinology* 33, 1378–1386.

Snigdha, S., de Rivera, C., Milgram, N. and Cotman, C. (2014) Exercise enhances memory consolidation in the aging brain. *Frontiers in Aging Neuroscience* 6, 3–15.

Adoption Preference Factors for Dogs in a Public Shelter in the Province of Palermo (Southern Italy) and Comparison with Adoption Rates of a Shelter in Northern Italy

Daniela Alberghina[1]*, Michele Panzera[1], Annalisa Macaluso[1], Pierluigi Raffo[2] and Annamaria Passantino[1]

[1]Department of Veterinary Science, University of Messina, Polo Universitario dell'Annunziata, Messina, Italy; [2]Associazione Arcadia Onlus, Località Lavini, Rovereto, Italy

Conflict of interest: The authors declare no conflict of interest.

Keywords: dog, public shelter, adoption rate, Southern Italy

Introduction

In Italy, each year approximately 150,000 dogs are taken into shelters. Only a small fraction is adopted, especially in the region of Sicily (southern Italy), where a large population of stray dogs exists. The goals of the study were: (i) to evaluate characteristics of adopted dogs in a public shelter in Sicily; and (ii) to compare data of population and adoption rate with those of a shelter in northern Italy.

Materials and Methods

A retrospective analysis of adopters' preference was conducted by comparing parameters such as age of the dog, gender, coat colour and size to the total canine population in a shelter in Palermo between 2014 and 2016. Similar data were collected from a shelter in northern Italy, and compared by Poisson regression.

---

* Corresponding author: dalberghina@unime.it

Results

The most important characteristics affecting the adopters' choices were age and size; young dogs were preferred over older dogs (adoptions/totals ratio 0.98) and small dogs were preferred over larger dogs (adoptions/totals ratio 0.87). Data analysis showed that in the shelter in Palermo, the canine population significantly increased during the 3 years of evaluation ($p<0.05$) and adoption rates were significantly lower than in Trento's Province shelter ($p<0.05$).

Conclusion

This preliminary study offers a quantitative analysis of adopters' preferences that could be useful to assess some adoptability index for dogs. Moreover, the study highlights the critical situation of over-population in Sicily's public dog shelters. It is paramount to find methods of development and validation of behavioural assessment protocols to increase successful adoption rates.

Relation Between the Owner's Psychological Well-being and Pets' Behavioural Problems: A Portuguese Survey

Joana Antunes de Almeida[1]*, Diogo Morais[2] and Gonçalo da Graça Pereira[3,4]

[1]VEEC - Hospital Veterinário Central, Charneca de Caparica, Portugal;
[2]Escola de Psicologia e Ciências da Vida (EPCV/ULHT), Lisboa, Portugal;
[3]CPCA - Centro para Conhecimento Animal, Algés, Portugal; [4]Escola Agrária de Elvas – Instituto Politécnico de Portalegre, Elvas, Portugal

Conflict of interest: The authors declare no conflict of interest.

Keywords: human–animal bond, depression, anxiety, stress, behavioural problems

Introduction

Owners report that at times their psychological well-being is affected by their pets' behaviour problems. Moreover, some owners may report that their own mental conditions may affect their pets. The aim of this study was to investigate possible correlation between owners' psychological well-being (depression, anxiety and stress) and the behaviour problems in their pets.

Material and Methods

A survey based study investigated a group of owners whose pets have behaviour problems (P) and owners of pets without behaviour problems (C). Questionnaires included questions regarding the animal, its environment and the owner's well-being using the Depression, Anxiety and Stress Scale.

* Corresponding author: joana.antunes.almeida@gmail.com

Results

Seventy-eight questionnaires were completed. Pets with behaviour problems appear to have owners with higher stress levels (t (76)=2.235; p=0.028). Also, castrated pets with behaviour problems have owners with higher depression levels (F (3.77)=4.200; p=0.044). There are significantly more married owners with pets that have behaviour problems when compared to single ones (χ^2 (4)=9.681; p=0.046). Finally, there is a tendency for cat owners to be more stressed than dog owners.

Conclusion

The result indicates a possible correlation between behaviour problems in pets and owners with reduced mental well-being. More studies are needed to confirm this putative correlation and, possibly, identify cause and effect.

The Association of Multiple Clinical Signs to Determine if a Cat Displays Either Urine Marking or Latrine Behaviour

ANA MARIA BARCELOS*, KEVIN MCPEAKE, NADJA AFFENZELLER AND DANIEL MILLS

School of Life Sciences, University of Lincoln, Lincoln, UK

Conflict of interest: The authors declare no conflict of interest.

Keywords: clinical signs, feline, house-soiling, latrine behaviour, urine marking

Introduction

In veterinary behavioural medicine, when presented with feline house-soiling problems it is important (but often challenging) to differentiate urine marking and latrine behaviour as this guides treatment recommendations. This study aimed to evaluate several aspects of urinary soiling behaviour in the home by cats to help differentiate marking and latrine behaviour.

Materials and Methods

An online questionnaire of 55 questions concerning the behaviour of cats observed when depositing urine with and without house-soiling (latrine or marking behaviour) problems was answered by 250 owners from 17 countries. Among those described as having house-soiling issues, at least one veterinary behaviourist assessed each individual questionnaire to determine whether the problem was related to marking or latrine behaviour. A consensus of three behaviourists was used where there was doubt. All described behaviours related to the soiling

* Corresponding author: anabhx@hotmail.com

©S. Denenberg 2017. *Proceedings of the 11th International Veterinary Behaviour Meeting* (ed. S. Denenberg)

act were analysed in order to identify patterns and associations among them. Five classic clinical signs (behaviours associated with the soiling act) were investigated for their association with the diagnosis of marking versus latrine behaviour in the house-soiling group: the cat's posture (stand vs squat), the surface chosen (vertical vs horizontal), the amount of urine (small vs medium vs large), the object selected (vertical vs horizontal) and the presence or absence of behaviours indicating covering the soiled area up afterwards. Sensitivity (i.e. occurrence of the classic sign in a case diagnosed with the relevant condition, such as standing to mark) and specificity (absence of the classic sign among those without the relevant condition, i.e. proportion not standing among latrine related cases) of individual clinical signs were also calculated.

Results

Classic latrine related signs were more sensitive than specific, but the converse was true of the classic marking signs. If known, the posture, the surface, volume of urine produced and the substrate (object) each had a sensitivity for the diagnosis of latrine related issue of more than 90%, but covering behaviour towards the elimination had a sensitivity of less than 80% for this type of problem; by contrast the classic signs of marking all had a sensitivity for this problem of less than 90%, with a small volume of urine having a sensitivity of less than 25%. If known, posture, surface, and substrate, all had similar specificities for the two conditions and were above 90%. Volume of urine and covering behaviour had a specificity of less than 90% for both conditions.

Conclusion

Some behaviours associated with the soiling act are more reliable than others to determine if a cat displays either urine marking or latrine behaviour, but a single sign should never be considered pathognomonic (sufficient) for the problem diagnosis.

Mirror Reflection or Real Image: Does Past Mirror Experience Influence a Dog's Use of a Mirror?

Megumi Fukuzawa* and Satomi Igarashi

Nihon University, College of Bioresource Sciences, Fujisawa-shi, Kanagawa, Japan

Conflict of interest: The authors declare no conflict of interest

Keywords: behaviour, dogs, mirror, real image, reflection

Introduction

Our aim was to investigate two factors, namely the dog's response to a person's reflection in a mirror or to the real person positioned in the same apparent position as the reflection, and the time spent by the dog looking in the mirror before and after the person was present.

Material and Methods

Nine pet dogs participated and were divided into two groups: a mirror-experienced group (EG) and a no-experience group (NEG). The EG had participated in a mirror test before the study. Two types of mirrors, a whole mirror and a half mirror, were used. A familiar person was positioned on the far side of an opaque barrier when the whole mirror was presented, or behind a clear panel when the half mirror was presented. Both the time it took the dog to reach the person and the dog's behaviours while reaching the person were recorded. The test was performed 20 times for each type of mirror for each dog, and a score of more than 17 out of 20 times was set as the criterion indicating that the dog could use the mirror.

---

* Corresponding author: fukuzawa.megumi@nihon-u.ac.jp

Results

Only five out of nine dogs reached the criterion in the half mirror session. All dogs achieved the criterion in the whole mirror session. The time taken to reach the person differed between dogs in the two mirror type sessions. The time spent looking in the mirrors differed only within the EG.

Conclusion

These results suggested that experience with a mirror affected the dogs' ability to solve mirror tasks.

Association Between Puppy Classes and Adult Behaviour of the Dog

Ángela González-Martínez[1]*, María Fuencisla Martínez[1], Maruska Suárez[2], Germán Santamarina[2], Belén Rosado[3], Isabel Luño[3], Sylvia García-Belenguer[3], Jorge Palacio[3], Ainara Villegas[3], Luis Felipe de la Cruz[4] and Francisco Javier Diéguez Casalta[2]

[1]Hospital Veterinario Universitario Rof Codina, Universidad de Santiago de Compostela, Lugo, Spain; [2]Departamento de Anatomía, Producción Animal y Ciencias Clínicas Veterinarias, Universidad de Santiago de Compostela, Lugo, Spain; [3]Departamento de Patología Animal, Facultad de Veterinaria, Universidad de Zaragoza, Zaragoza, Spain; [4]Departamento de Fisiología, Universidad de Santiago de Compostela, Lugo, Spain

Conflict of interest: The authors declare no conflict of interest.

Keywords: dogs, puppy class, behaviour problems, behaviour

Introduction

A puppy's early environment may have a profound effect on its future behaviour, making appropriate socialization and habituation during the early weeks of life essential for lifelong well-being (Sforzini *et al.*, 2009). Nevertheless, the literature shows opposite effects between puppy classes and adulthood behaviour (Seksel *et al.*, 1999; Batt *et al.*, 2008; Blackwell *et al.*, 2013; Kutsumi *et al.*, 2013; Howell *et al.*, 2015). The aim of this study was to assess the effect of puppies' attendance at dog-training programmes on their behaviour later in adulthood.

* Corresponding author: angela_982@hotmail.com

©S. Denenberg 2017. *Proceedings of the 11th International Veterinary Behaviour Meeting* (ed. S. Denenberg)

Material and Methods

Eighty dogs were enrolled into this study. Thirty-two dogs had attended puppy classes at VTH Rof Codina, whereas the remaining 42 had not; these 42 were recruited through social networks or email. Descriptive data of the studied dog population are presented in Table 1. All dogs were evaluated using the C-BARQ 1 year after the completion of the puppy classes. Ordinal regression models were used to estimate the influence of puppy classes on the different behavioural traits assessed by the C-BARQ.

Results

The results indicated that dogs that had attended classes had more favourable scores for family dog aggression (p=0.04), trainability (p=0.02), non-social fear (p=0.02), excitability (p=0.05) and touch sensitivity (p=0.017). No relation was found between attending puppy classes and stranger-directed aggression, owner-directed aggression, dog-directed aggression, dog-directed fear, separation-related problems, chasing, stranger-directed fear, attachment/attention seeking or energy.

Conclusion

The study showed that attending puppy class is important for socialization with other puppies and people, which will have an impact on the dog's long-term behaviour.

Table 1. Descriptive analysis of the studied dog population.

Variable			Descriptives	
Puppy classes	Yes	32 (40%)	Before 3 months of age	15 (46.9%)
			After 3 months of age	17 (53.1%)
	No	48 (60%)		
Morphotype	Small	23 (28.8%)		
	Medium	37 (46.2%)		
	Large	20 (25%)		
Gender	Male	42 (52.5%)		
	Female	38 (47.5%)		
Mean age at acquisition (months)		3.48 (S.D. = 2.6)		
Place of acquisition	Shelter	40 (50%)		
	Other	40 (50%)		
Neutered	Yes	45 (43.8%)	Due to a behaviour problem	7 (15.5%)
			Other reason	38 (84.5%)
			Before 6 months of age	12 (26.7%)
			After 6 months of age	33 (73.3%)
	No	35 (56.3%)		

References

Batt, L., Batt, M., Baguley, J. and McGreevy, P. (2008) The effects of structured sessions for juvenile training and socialization on guide dog success and puppy-raiser participation. *Journal of Veterinary Behavior: Clinical Applications and Research* 3, 199–206.

Blackwell, E.J., Bradshaw, J.W. and Casey, R.A. (2013) Fear responses to noises in domestic dogs: prevalence, risk factors and co-occurrence with other fear related behaviour. *Applied Animal Behavior Science* 145, 15–25.

Howell, T.J., King, T. and Bennett, P.C. (2015) Puppy parties and beyond: the role of early age socialization practices on adult dog behavior. *Veterinary Medicine: Research & Reports* 6, 143–153.

Kutsumi, A., Nagasawa, M. and Ohtani, N. (2013) Importance of puppy training for future behavior of the dog. *Journal of Veterinary Medical Science* 75, 141–149.

Seksel, K., Mazurski, E.J. and Taylor, A. (1999) Puppy socialisation programs: short and long term behavioural effects. *Applied Animal Behavior Science* 62, 335–349.

Sforzini, E., Michelazzi, M., Spada, E., Ricci, C., Carenzi, C., Milani, S., Luzi, F. and Verga, M. (2009) Evaluation of young and adult dogs' reactivity. *Journal of Veterinary Behavior: Clinical Applications and Research* 4, 3–10.

The Importance of Pain as a Differential Diagnosis During a Behaviour Consultation

LYNN HEWISON[1]*, DAVE ELLSON[2], ALEX HAMILTON[3] AND KEVIN MCPEAKE[1]

[1]Animal Behaviour, Cognition and Welfare Group, School of Life Sciences, University of Lincoln, Lincoln, UK; [2]WildboreVETStop Ltd, Worksop, UK; [3]Willows Veterinary Centre & Referral Service, Solihull, UK

Conflict of interest: The authors declare no conflict of interest.

Keywords: pain, behaviour, dog, anxiety, fear, investigations

History and Presenting Signs

A 5-year-old neutered male Greyhound was referred to the Animal Behaviour Referral Clinic, University of Lincoln, for pulling backwards and bolting on walks. Prior to referral, the dog had not responded to treatment with a neutraceutical (Nutracalm®). At the time of the consultation, the dog's behaviour had worsened to the extent he was often refusing daily walks. During the consultation, the dog was observed to be stiff in his hind quarters and described by his owners as 'lacking stamina', 'less active after longer walks' and 'squatting to urinate'.

Diagnosis

After the assessment, the dog was diagnosed as fearful and anxious. However, the history and consultation process also suggested that pain was contributing to the behaviour and may, in fact, be the chief underlying cause.

---

* Corresponding author: lhewison@lincoln.ac.uk

Physical Examination

The dog returned to the referring veterinarian for further investigations including radiographs of his hips and lumbar spine which showed no specific pathology. However, as pathological changes on radiographs do not necessarily correlate with the pain experienced by an individual (Dobromylskyj *et al.*, 2000), a 4–6 week pain relief trial was also recommended.

The dog showed little improvement on a (3-week) non-steroidal anti-inflammatory (NSAID) trial but marked improvement within 24 hours of gabapentin being added. Based on the response to gabapentin and unremarkable radiographs, the patient was referred to a neurologist who diagnosed idiopathic neuropathic pain after further tests (including magnetic resonance imaging, cerebrospinal fluid analysis).

Management and Follow-up

The dog was maintained on gabapentin long term, which resulted in marked improvement in the dog's behaviour, even though minimal behaviour modification was implemented. The behaviour continues to improve 5 months post-referral.

Practical Applications

This case report demonstrates the importance of considering pain as a differential when an animal is referred for a behaviour problem. It also shows that a short trial with a NSAID may be insufficient to rule out pain, and therefore alternative classes of analgesia may have to be considered alongside further investigations.

References

Dobromylskyj, P., Flecknell, P.A., Lascelles, B.D., Livingston, A., Taylor, P. and Waterman-Pearson, A. (2000) Pain assessment. In: Flecknell, P.A. and Waterman-Pearson, A. (eds) *Pain Management in Animals*. W.B. Saunders, London, pp. 53–79.

Canine Electroencephalograph Transients in the Sigma Range Relate to Memory

Ivaylo Borislavov Iotchev[1], Anna Vargáné Kis[2], Róbert Bódizs[3,4], Gilles van Luijtelaar[5] and Enikő Kubinyi[1]*

[1]Department of Ethology, Eötvös Loránd University, Budapest, Hungary; [2]Institute of Cognitive Neuroscience and Psychology, Hungarian Academy of Sciences, Budapest, Hungary; [3]Institute of Behavioural Sciences, Semmelweis University, Budapest, Hungary; [4]Department of General Psychology, Pázmány Péter Catholic University, Budapest, Hungary; [5]Donders Centre of Cognition, Radboud University Nijmegen, The Netherlands

Funding: This study was funded by a grant from the European Research Council (ERC) under the European Union's Horizon 2020 research and innovation programme (Grant Agreement No. 680040) and from the Bolyai Foundation of the Hungarian Academy of Sciences.

Conflict of interest: The authors declare no conflict of interest.

Keywords: behaviour, EEG, polysomnography, dogs, learning, sleep spindle

Introduction

Sleep spindles are brief bursts of cortico-thalamo-cortical activity, which are visible in the cortex as transient oscillations in the sigma range (usually defined as 12–14 Hz or 9–16 Hz). In humans and rodents, they have been associated with sleep-dependent memory consolidation and sleep stability, as well as mechanisms that could plausibly explain these relationships. In addition, sleep spindles were also found to change in occurrence, frequency, amplitude, and duration in response to age, sex and psychiatric conditions. Although the dog represents a promising model of human (social) behaviour, brain function, and ageing, spindle

* Corresponding author: eniko.kubinyi@ttk.elte.hu

©S. Denenberg 2017. *Proceedings of the 11th International Veterinary Behaviour Meeting* (ed. S. Denenberg)

analogue activity has only been described and never systematically quantified and related to function in this species.

Material and Methods

In the present study, we used an adjusted version of a detection method previously tested in human children and a data set of electroencephalograph (EEG) and memory measurements obtained from dogs, to test the predictive validity of automatic detections in the dog as a potential model of the human spindle.

Results

We found that the density of EEG transients in the 9–16 Hz range and non-REM sleep phase reflect the same relationship to memory and sexual dimorphism as in humans. However, age-related effects were only marginally significant, and could have been masked by the sample size and large sex differences in spindle density.

Conclusion

We conclude that automatic detections in the 9–16 Hz range are promising analogues of human spindles and can potentially widen the utility of non-invasive polysomnographic methods in this species.

Noise Sensitivities in Dogs: An Exploration of Signs in Dogs With and Without Musculoskeletal Pain

ANA LUISA LOPES FAGUNDES[1], LYNN HEWISON[2], KEVIN J. MCPEAKE[2]*, HELEN ZULCH[2] AND DANIEL MILLS[2]

[1]Centro Universitário de Belo Horizonte, Belo Horizonte, Minas Gerais, Brazil; [2]Animal Behaviour Clinic, School of Life Sciences, University of Lincoln, Lincoln, UK

Conflict of interest: The authors declare no conflict of interest.

Keywords: anxiety, behaviour, dog, fear, noise sensitivity, pain

Introduction

Noise sensitivity is a common behaviour problem in dogs, the onset of which may be associated with a range of medical conditions. In humans, there is a well-established but complex relationship between painful conditions and the development of fear-related avoidance responses, however, while there are reasons for believing that a relationship exists between noise sensitivity and pain in dogs, this does not appear to have been investigated. The aim of the current study was to use qualitative thematic analysis to compare the case histories of dogs presented for noise sensitivity where an identifiable pain focus was present or absent.

Material and Methods

Data were extracted from the clinical records of 20 cases of dogs presenting with noise sensitivity seen by clinical animal behaviourists at the Animal Behaviour Clinic, University of Lincoln (10 'clinical cases' with a focus on musculoskeletal pain; 10 'control cases' with no identified pain focus).

* Corresponding author: kmcpeake@lincoln.ac.uk

Results

Loud noises were a common trigger of a fear reaction in all cases, but ubiquitous among the 'clinical cases'. The age of onset of noise sensitivity was on average nearly 4 years later in 'clinical cases'. In 'clinical cases', noise sensitivities often involved broad generalisation to associated environmental cues, and in addition, problematic behaviour with other dogs was also more commonly reported.

Conclusion

Veterinarians and behaviourists should carefully assess the potential for a pain-related problem in dogs with noise sensitivities presenting with these characteristics. These results should be considered suggestive and not diagnostic, and future work should look to quantify the size of these potential effects.

Evaluation of Urine Cortisol:Creatinine Ratio in Dogs with Separation Anxiety

JULIA MILLER[1]*, BARBARA FABICH[2], KAROLINA OWSIŃSKA[3], KATARZYNA WIĘCEK[3] AND ANNA CHEŁMOŃSKA-SOYTA[1]

[1]Department of Immunology, Pathophysiology and Veterinary Preventive Medicine, Wrocław University of Environmental and Life Sciences, Wrocław, Poland; [2]Private practice; [3]Student at the Faculty of Veterinary Medicine, Wrocław University of Environmental and Life Sciences, Wrocław, Poland

Funding: The study was funded by Faculty of Veterinary Medicine, Wrocław University of Environmental and Life Sciences.

Conflict of interest: The authors declare no conflict of interest.

Keywords: separation anxiety, stress, cortisol, dogs

Introduction

Separation anxiety is a common cause of behavioural consultations. Destructive behaviours, house-soiling or vocalisation can be observed by the owners after return or reported by neighbours. However, there is a wide range of differential diagnoses for each of these behaviours. The aim of the study was to evaluate the usefulness of the urine cortisol:creatinine ratio as an assessment tool in these cases.

Materials and Methods

Questionnaires and observation of recorded behaviours were used to make a diagnosis. Seven affected dogs (three females and four males, mean age 3.85±1.68) and ten control dogs (four females and six males; mean age 5.7±3.2) were included. Urine samples were obtained after at least a 2-hour absence (test day) and at a

* Corresponding author: julia.miller@upwr.edu.pl

similar time on a day the dog spent together with the owner (control day). The cortisol:creatinine ratio was measured by chemiluminescence. The results were analysed by Mann–Whitney U Test.

Results

The average cortisol:creatinine ratio in the urine samples obtained on test days from affected dogs and control dogs was 31.58 and 14.8, respectively (p<0.01). The most distinct difference (p<0.001) was seen when comparing the differences in the urine cortisol:creatinine ratio between the control day and test day (13.1 and 1.9 for the affected and control group, respectively).

Conclusion

Urine cortisol:creatinine ratio was found to be helpful in evaluating the stress re-action in dogs suffering from separation anxiety. This study is ongoing to evaluate the urine cortisol:creatinine ratio in a larger group of affected dogs. We also plan to expand our research and compare the cortisol levels in different samples (e.g. saliva or blood samples).

Effects of Providing Two Kinds of Doghouses in Different Sized Pens on the Behaviour of Shelter Dogs

Simona Normando[1]*, Barbara Contiero[2], Alessandra Azzolini[2] and Rebecca Ricci[2]

[1]Dept BCA, Padua University, Legnaro (PD), Italy; [2]Dept MAPS, Padua University, Legnaro (PD), Italy

Conflict of interest: The authors declare no conflict of interest.

Keywords: behaviour, dogs, space allowance, doghouse

Introduction

Large space and a doghouse can be enriching for shelter dogs; however, the cost of housing is higher. Investigating this hypothesis is important to evaluate the cost:benefit ratio. The aim of this study was to assess the effects of providing two kinds of doghouses in different size pens on the behaviour of shelter dogs.

Material and Methods

Five pairs of shelter dogs (seven castrated males, three spayed females, aged 3–12 years) were included. Following a corrected Latin square design, dogs were exposed to four 1-week-long experimental conditions, created by combining two pen sizes (9 vs 18 square meters) with the provision of a flat topped either accessible or inaccessible doghouse (i.e. in either case dogs could be on top of the doghouse, but could or could not be inside it). Dogs were observed by instantaneous scan sampling every 20 seconds for 20 minutes/day/pair for 4 days per condition and by continuous focal animal sampling for the first 10 minutes of each new condition. Data were analysed using SAS PROC GEN MOD and PROC MIXED, respectively.

---

* Corresponding author: simona.normando@unipd.it

Results

The risk of showing self-grooming (p<0.001) was lower in the larger pen size. Also, the combination of a larger pen and accessible doghouse reduced the risk of self-grooming (p=0.04), as compared to the other combinations. Larger pen size was also associated with increased activity (p=0.002), sniffing (p=0.03), social interactions (p=0.02), and decreased lip-licking and yawning (p=0.0073) for the first 10 minutes of housing.

Conclusion

Larger pen size and access to a doghouse appear to have a positive effect on the behaviour of shelter dogs.

Circadian Distribution and Characterisation of Social Behaviour in a Group of Domestic Donkeys (*Equus asinus*)

SIMONA NORMANDO*, PAOLO MONGILLO, FRANCESCA VISENTIN, RAFFAELLA AMINA MOSANER AND LIETA MARINELLI

Dept BCA, Padua University, Legnaro (PD), Italy

Conflict of interest: The authors declare no conflict of interest.

Keywords: donkeys, social behaviour, circadian rhythm

Introduction

Following a decrease in number, the domestic donkey population in Italy is rising, due mainly to their involvement in meat/milk production and Animal Assisted Interventions. However, to date only little scientific literature exist on this species' behaviour. The aim of this study was to investigate intra-species affiliative social behaviour (e.g. allo-grooming, proximity maintenance, contact maintenance) in donkeys.

Materials and Methods

Thirteen donkeys (10 females, 3 geldings, aged 3–13 years) stabled in a 480 m² pen with a 45 m² stable were observed. In phase I, donkeys' behaviours were recorded from videos using an instantaneous scan sampling method for 20 minutes every other hour for 3 non-consecutive days to assess circadian distribution of social behaviour. In phase II, behaviours were recorded using a continuous focal animal sampling method for 4 days during the hour of the day in which most of the social behaviours had taken place during phase I, to define intra-species social interactions.

---

* Corresponding author: simona.normando@unipd.it

Results

Donkeys were involved in affiliative intra-specific social behaviour in 14.0% of the scans in which they were visible. There were significant differences in the occurrence of social behaviour among hours of the day ($p < 0.00001$, Friedman test) with most social behaviours taking place between 08:00 and 09:00. The sociogram showed five often-interacting dyads, three composed of unrelated females and two by mother and offspring, characterised mostly by proximity maintenance.

Conclusion

In this study, affiliative social behaviour of donkeys was expressed mainly by dyadic proximity maintenance, and followed a circadian rhythm.

Effect of Meal Composition on Tryptophan:Large Neutral Amino Acids Ratio in Dogs

Asahi Ogi*, Beatrice Torracca, Chiara Mariti, Lucia Casini and Angelo Gazzano

Università di Pisa, Dipartimento di Scienze Veterinarie, Pisa, Italy

Conflict of interest: The authors declare no conflict of interest.

Keywords: tryptophan, LNAAs, serotonin, behaviour, meal, dog, HPLC

Introduction

Tryptophan is involved in the synthesis of serotonin and melatonin, and it competes with the other large neutral amino acids for uptake into the brain (Fernstrom, 2013). The aim of this study was to assess the impact of meal composition on the plasma ratio between tryptophan and five other large neutral amino acids (5LNAAs): isoleucine, leucine, valine, tyrosine and phenylalanine.

Material and Methods

This study included five female Labrador Retrievers (two spayed, 8.6 ± 3.8 years old) from the same bloodline and usually fed the same commercial food once a day. Each dog received a meal of puffed rice, minced meat and olive oil (M1) for 1 day. Then the dogs were fed their normal diet for 30 days. Finally, the dogs received a meal of puffed rice and olive oil (M2), with no meat, for 1 day. A second meal was administered in the evening to balance the energy intake, and both diets were isoenergetic. Blood was collected immediately before feeding (t0) and after 2, 4, 6, 8 and 10 hours. Plasma samples were used for HPLC quantification of tryptophan and 5LNAAs (Wu and Meininger, 2008). At each sampling time, their levels and ratio after M1 and M2 were compared using Wilcoxon rank-sum test ($p < 0.05$).

---

* Corresponding author: asahi.ogi@vet.unipi.it

©S. Denenberg 2017. *Proceedings of the 11th International Veterinary Behaviour Meeting* (ed. S. Denenberg)

Results

Tryptophan and phenylalanine concentrations showed no significant difference between M1 and M2 samples. Isoleucine, leucine, valine and tyrosine plasma concentrations were lower after M2. Tryptophan:5LNAAs ratio was higher following the meal without meat (M2) at all sampling times except t0. This trend was a statistically significant difference at 2 (median: 0.206 vs 0.311), 4 (median: 0.217 vs 0.345) and 10 (median: 0.242 vs 0.289) hours following the meal.

Discussion

These findings suggest that meal composition plays a vital role in the tryptophan bioavailability. However, further studies, observing behavioural changes and assessing serotonin and melatonin levels, are required to evaluate the impact of tryptophan:LNAAs ratio on dog behaviour.

References

Fernstrom, J.D. (2013) Large neutral amino acids: dietary effects on brain neurochemistry and function. *Amino Acids* 45(3), 419–430.

Wu, G. and Meininger, C.J. (2008) Analysis of citrulline, arginine, and methylarginines using high-performance liquid chromatography. In: Cadenas, E. and Packer, L. (eds) *Methods in Enzymology*, Vol. 440. Elsevier Academic Press Inc., San Diego, California, pp. 177–189.

Differences in Management of Dogs and Cats Living in the Same Household

Laura Menchetti[1], Silvia Calipari[1], Alice Catanzaro[2], Chiara Mariti[3], Angelo Gazzano[3] and Silvana Diverio[1]*

[1]Laboratory of Ethology and Animal Welfare (LEBA), Department of Veterinary Medicine, Perugia University, Italy; [2]Veterinary Consultant, London, UK; [3]Department of Veterinary Sciences, University of Pisa, Italy

Conflict of interest: The authors declare no conflict of interest.

Keywords: dog, cat, management, multi-pet households, owner–pet relationship

Introduction

Despite the popular belief about their incompatibility, dogs and cats commonly cohabitate in the same household. The present study investigates whether owner management choices and relational approach are different for the two species.

Material and Methods

Data were collected from 1270 questionnaires filled out by owners of dogs and cats living in the same household. Owners and pets, demographic characteristics, pet management and pet–owner relationship were analysed by Univariate and Cluster Analysis.

Results

Neutering rate was higher in cats (89%) than in dogs (57%; $p<0.001$) and when multiple pets were owned ($p<0.05$). We found agreement in living habits between

* Corresponding author: silvana.diverio@unipg.it

dogs and cats living in the same household only in 19.4% of owners. Most dogs lived indoors (53%; p<0.001), while most of the cats lived both indoors and outdoors (48%; p<0.001). Several dogs (35%) and cats (48%) slept free in the home (p<0.001). Most of the participants (59%) described the dog–owner relationship as similar to the cat–owner relationship: as loving (30%) or friendly (28%). We found concordance in sleeping habits between dogs and cats in almost half of the owners (43%). Pets' sleeping habits differed in relation to pet–owner relationship (p<0.01), with a higher proportion of 'Loving relationship' owners allowing their dog to sleep free in the home (41%) and their cat on the bed (23%).

Conclusion

The relationship that owners build with their pets did not diversify depending on the species. However, the management style often showed dissimilarities, especially in living and sleeping habits, which seem to be correlated to the relationship itself.

Acupuncture Influence on Reduction of Stress Signs in Kennelled Dogs

BRUNO SOUSA[1]*, SARA FRAGOSO[2], LUÍS RESENDE[3], MARIA PAULA MIRANDA[4], GLÓRIA TANECO[5], INÊS VIEGAS[6] AND GONÇALO DA GRAÇA PEREIRA[2,7,8]

[1]*Universidade Lusófona de Humanidades e Tecnologias, Lisbon, Portugal;* [2]*Centro Para o Conhecimento Animal, Algés, Portugal;* [3]*ChiVet – Acupuntura Veterinária, Lisbon, Portugal;* [4]*Clínica Veterinária Nova Vetcarnide, Lisbon, Portugal;* [5]*Centro Clínico Veterinário de Aires, Lisbon, Portugal;* [6]*Petable, Lisbon, Portugal;* [7]*Escola Superior Agrária de Elvas, Instituto Politécnico de Portalegre, Elvas, Portugal;* [8]*Centro de Estudos de Ciência Animal, Instituto de Ciências, Tecnologias e Agroambiente, Universidade do Porto, Porto, Portugal*

Conflict of interest: The authors declare no conflict of interest.

Keywords: acupuncture, dog, kennel, stress, animal welfare

Introduction

Many acupuncture modalities are used to produce therapeutic effects. In this research, we choose pharmacopuncture: an injection of medicinal materials on acupuncture points, to combine the benefits of both (Kim and Kang, 2010). It has the additional advantage of sparing time, limiting the need to restrain the animal during the acupuncture session (15–20 minutes), which is difficult for kennelled dogs. The objective is to reduce fear and anxiety, improve the relationship between dogs and kennel staff, and consequently improve welfare, without behavioural modification.

* Corresponding author: bmasousa.mv@gmail.com

©S. Denenberg 2017. *Proceedings of the 11th International Veterinary Behaviour Meeting* (ed. S. Denenberg)

Material and Methods

Forty-five kennel dogs, of varying age and sex, were included this study. Each dog was filmed and its behaviour was assessed and recorded, based on an ethogram, for 4 months. The dogs were divided into three groups. Dogs in group A were taken into the treatment room, received pharmacopuncture and were returned to their kennels. Dogs in group B had similar protocol without pharmacopuncture. Dogs in group C were left in their kennels without pharmacopuncture. The observer was blind during the study. The pharmacopuncture was administered by the same three veterinarians. Once a week, during the last 2 months, dogs in group A were stimulated using the following acupoints, with cobamamide/cyanocobalamin injection: *Yin Tang, Da-feng-men, Tian men, An-Shen, Nei-guan* and *Shen-men*.

Results

No significant differences ($p \leq 0,05$) were found following pharmacopuncture (except for 1 variable out of 12), without behavioural modification, on diminishing stress signs.

Conclusion

The outcome of this study could not prove the effects or efficacy of acupuncture therapy on reducing stress signs on kennelled dogs.

Reference

Kim, J. and Kang, D.A. (2010) Descriptive statistical approach to the Korean pharmacopuncture therapy. *Journal of Acupuncture and Meridian Studies* 3, 141–149.

Serum Leptin and Ghrelin Levels and their Relationship with Serum Cortisol, Thyroid Hormones, Lipids, Homocysteine and Folic Acid in Dogs with Compulsive Tail Chasing*

Ebru Yalcin[1][†], Zeki Yilmaz[1] and Yesim Ozarda[2]

[1]Department of Internal Medicine, Faculty of Veterinary Medicine, Uludag University, Bursa, Turkey; [2]Department of Biochemistry, Medical School, Uludag University, Bursa, Turkey

Conflict of interest: The authors declare no conflict of interest.

Keywords: compulsive tail chasing, behaviour, leptin, ghrelin, cortisol, thyroid

Introduction

Compulsive disorder (CD) is a neuropsychiatric disorder in humans and animals, and can be anxiety-related. Compulsive disorders include excessive tail chasing, light/shadow chasing and flank sucking (Overall and Dunham, 2002). The aim of this study was to investigate serum leptin and ghrelin levels, and their relationship with circulating cortisol, thyroid hormones, lipids, homocysteine (Hcy) and folic acid in dogs with compulsive tail chasing (CTC).

---

* This study has been published in full in *Journal of Veterinary Medicine*, Kafkas University, at: http://vetdergi.kafkas.edu.tr/extdocs/2017_2/227-232.pdf
† Corresponding author: yalcine@uludag.edu.tr

Material and Methods

Fifteen dogs with CTC and 15 healthy control dogs of various weights, breeds, ages and sexes were used. Diagnosis of CTC was based on the dog's behavioural history, clinical signs and ruling out possible medical reasons. Blood samples were collected from all dogs.

Results

Leptin was higher in CTC dogs compared to control (8.3±0.9 ng/ml vs 1.7±0.2 ng/ml, p<0.001). Ghrelin level was lower in CTC dogs vs control (74±7 pg/ml vs 144±41 pg/ml, p<0.05). Serum cortisol, lipid (cholesterol, phospholipids and NEFA) and Hcy levels increased (p<0.05), whereas serum folic acid decreased (p<0.001) in CTC dogs compared to control dogs. Serum ghrelin level correlated negatively with cholesterol (p<0.05), and serum leptin level correlated positively with cholesterol, fT4 and phospholipids (p<0.05).

Discussion

The results of serum leptin in this study showed similarity to those of a human study (Emul *et al.*, 2007). Serum leptin and ghrelin might have a role in regulating circulating lipids (total cholesterol, HDL-C, LDL-C and phospholipids) in dogs with CTC. Serum leptin correlated positively only with serum fT4 levels amongst thyroid hormones measured in this study. The presence of hyperhomocysteinaemia and low serum folate in CTC dogs is similar to reports in human studies on OCD (Atmaca *et al.*, 2005; Turksoy *et al.*, 2014).

Conclusions

Serum leptin and ghrelin levels may bring up a new perspective on our understanding of the pathophysiological mechanisms associated with CTC. Serum levels of both hormones may be associated with serum levels of lipids and free T4.

References

Atmaca, M., Tezcan, E.., Kuloglu, M., Kirtas, O. and Ustundag, B. (2005) Serum folate and homocysteine levels in patients with obsessive-compulsive disorder. *Psychiatry and Clinical Neurosciences* 59, 616–620.

Emul, H.M., Serteser, M., Kurt, E., Ozbulut, O., Guler, O. and Gecici, O. (2007) Ghrelin and leptin levels in patients with obsessive-compulsive disorder. *Progress in Neuro-psychopharmacology & Biological Psychiatry* 31, 1270–1274.

Overall, K.L. and Dunham, A.E. (2002) Clinical features and outcome in dogs and cats with obsessive-compulsive disorder: 126 cases (1989–2000). *Journal of the American Veterinary Medical Association* 221, 1445–1452.

Turksoy, N., Bilici, R., Yalciner, A., Ozdemir, Y.O., Ornek, I., Tufan, A.E. and Kara, A. (2014) Vitamin B12, folate, and homocysteine levels in patients with obsessive compulsive disorder. *Neuropsychiatric Disease Treatment* 10, 1671–1675.

A Social Entrepreneurship Project on Animal Welfare: 'Pretty Paws' Board Game

EBRU YALCIN*

Department of Internal Medicine, Faculty of Veterinary Medicine, Uludag University, Bursa, Turkey

Conflict of interest: The author declares no conflict of interest.

Keywords: behaviour, animal welfare, social entrepreneurship, board game

Introduction

Lack of knowledge about pets' health, behaviour and welfare can compromise the pet's well-being. Children appear to be especially susceptible to this. There is a need to educate children about pets, including health, behaviour, nutrition, immunisation, welfare and management. One way to educate children is to introduce the knowledge through games. The aim of this project was to evaluate the effect of a board game on educating children about pets.

Materials and Methods

Fifty children between the ages of 7 and 12 years, in local primary schools, were given a board game. Two or more children can play the game. The teachers evaluated the effect on the children and their knowledge. The game included topics such as the veterinarian, veterinary clinic, shelter, pets' behaviour and socialisation. The game consisted of a board with curved instructions on it with pawns and game cards.

* yalcine@uludag.edu.tr

©S. Denenberg 2017. *Proceedings of the 11th International Veterinary Behaviour Meeting* (ed. S. Denenberg)

Results

Our preliminary observation indicates that children who played this board game have given positive feedback about the pet's welfare, health and behaviour. This project is still ongoing.

Conclusion

Providing knowledge to children through games appears to have beneficial effects. Children were able to understand the game and learn the facts. More studies are needed to evaluate further the full effect of the game.